Cambridge IGCSE®

Chemistry

STUDY AND REVISION GUIDE

David Besser

HODDER
EDUCATION
AN HACHETTE UK COMPANY

Author's dedication

To Martha, Sarah and Joseph. Thank you for everything.

Hachette UK's policy is to use papers that are natural, renewable and recyclable products and made from wood grown in sustainable forests. The logging and manufacturing processes are expected to conform to the environmental regulations of the country of origin.

Orders: please contact Bookpoint Ltd, 130 Park Drive, Milton Park, Abingdon, Oxon OX14 4SE. Telephone: (44) 01235 827720. Fax: (44) 01235 400454. Email: education@bookpoint.co.uk Lines are open from 9 a.m. to 5 p.m., Monday to Saturday, with a 24-hour message answering service. You can also order through our website: www.hoddereducation.com

ISBN 978 1471 894 602

® IGCSE is the registered trademark of Cambridge International Examinations. The questions, example answers, marks awarded and/or comments that appear in this book were written by the author. In examination, the way marks would be awarded to answers like these may be different.

This book has not been through the Cambridge endorsement process.

© David Besser 2017

First published in 2017 by
Hodder Education,
An Hachette UK Company
Carmelite House
50 Victoria Embankment
London EC4Y 0DZ

www.hoddereducation.com

Impression number   10 9 8 7 6 5 4 3 2 1

Year   2019 2018 2017

Cover photo © fox17 – Fotolia
Illustrations by Integra Software Services Pvt. Ltd
Typeset in ITC Galliard Std Roman 11/13 by Integra Software Services Pvt. Ltd., Pondicherry, India
Printed in Spain

A catalogue record for this title is available from the British Library.

Contents

REVISED

Introduction

Welcome to the Cambridge IGCSE® Chemistry Study and Revision Guide. This book has been written to help you revise everything you need to know for your Chemistry exam. Following the Chemistry syllabus, it covers all the key content as well as sample questions and answers, practice questions and examiner tips to help you learn how to answer questions and to check your understanding.

● How to use this book

Key objectives
The key skills and knowledge covered in the chapter. You can also use this as a checklist to track your progress.

● Key terms

Definitions of key terms you need to know from the syllabus.

● Sample exam-style questions

Exam-style questions for you to think about.

Student's answers

Typical student answers to see how the question might have been answered.

Examiner's comments

Feedback from an examiner showing what was good, and what could be improved.

Examiner's tips
Advice to help you give the perfect answer.

Common errors

● Mistakes students often make and how to avoid them.

● Extended

Content for the extended syllabus is shaded green.

Exam-style questions
Practice questions for you to answer so that you can see what you have learned.

Cambridge IGCSE Chemistry Study and Revision Guide © David Besser

● How to revise

This book is not intended to give detailed information about the chapters you are required to study for the IGCSE Chemistry course. Instead it is meant to give concise information concerning the things that you are likely to come across in your examinations. You have probably been using a more detailed textbook over the two years of your course. This book is intended for use over the six weeks just before the examinations.

No two people revise in the same way. It would be foolish to give precise instructions to anyone about how they should prepare for examinations. However, I intend to make some suggestions about the different approaches that are available, so that you can choose the methods that are most suitable for you.

The only thing I would strongly recommend about revision is that it should involve writing as well as reading. Those who read through notes or books as their only means of revising often find that they become distracted and start thinking about other things. Writing things down helps you to focus on what you are trying to learn.

Another way to help you learn is to highlight key words and phrases that you wish to draw attention to. Highlighting makes you focus on things that you may have had problems with up to now.

After highlighting, you could rewrite the highlighted parts, leaving out the less important parts. It may also be a good idea to leave out those parts that you already know. If you know that transition elements are all metals, there is no point in writing it down, because you do not need to revise it. Just focus on the parts that you are not so familiar with. Your notes will be more concise and more personal than the information in the book. You may prefer to write down the information in a more eye-catching form, such as in a diagram. The important thing is that it is personal to you and helps your revision.

When you have made notes of this type, try writing them out. Again just focus on the important key words and phrases. When you can write them out without looking at your notes, you may be confident that you have learned this particular chapter. You can test yourself by answering the 'Exam-style questions' sections of the book as well as looking at past paper questions or questions you have attempted during the course. If you come across a problem, always go back to your notes and other sources of information. There is no need to attempt timed questions under exam conditions until much nearer the exam.

● Exam technique

The most important pieces of advice that anyone can give about how to approach an exam are not new. However, it is well worth repeating them, because the points made below are still major causes of avoidable losses of marks.

- **Write clearly**. If examiners cannot read what you have written, they cannot award marks, despite their best efforts to decipher your work. In this technological age, when handwriting is not used as much as it was in the past, it is essential that your answers can be read.
- **Use correct spellings**. Technical terms used in Chemistry must be written correctly. Incorrect spellings do more than create a bad impression of a candidate's work. If a word is incorrectly spelt, it often changes the whole

v

meaning of the word. An example is using the word 'alkane' instead of 'alkene'. This is not only a spelling mistake, but it replaces one word with an entirely different word which has an entirely different meaning.

- **Read the questions carefully**. Do not just glance at a question and pick out a few words. Read the whole question and when you have read it, read it again. A question may look like another one you have seen during your revision, but if you read it more carefully, you may realise that there are differences which mean that a different approach is required to answer the question.
- **Check your answers**. When you have finished each question, read through it to make sure it makes sense and that it answers the question.
- **Do not spend too long on any questions**. If you spend too long on some questions, you may find you do not have the necessary amount of time to answer some of the others. It is important to answer all the questions.

Some common phrases that you will see in questions, and their meanings, include:

- **Define the term/what is meant by the term** means give a definition of a word or phrase which only applies to that word or phrase. For example, 'Define the term isotope' means give a statement that tells someone exactly what an isotope is. Definitions are found in the 'Key terms' section at the start of each chapter.
- **State** means give a brief statement. No explanation of the statement is required. For example, 'State the name of the acid that is used to make magnesium sulfate'.
- **Explain/give a reason or reasons**. This sometimes follows the command word 'state', i.e. 'state and explain' or 'state and give a reason'. This means you should give a piece of information followed by a brief explanation of why you chose this information.
- **Outline** means a brief description is required.
- **Predict** means you are meant to make a prediction, not based on any knowledge that you have remembered, but by making a logical connection between other pieces of information referred to in the question.
- **Deduce** also means you are not required to give an answer based on what you have remembered, but to suggest a logical connection based on information in the question.
- **Suggest** may mean there is more than one possible answer. It may also mean you are required to apply your knowledge of Chemistry to a 'novel' situation, e.g. an experiment or a reaction you have not come across before. In such examples, you will be supplied with sufficient information to make a reasonable suggestion.
- **Calculate/determine** means carry out a calculation based on data that is provided.
- **How would you ...?** means give a brief description of an experiment that you would carry out. Many exam candidates answer this type of question with too much theoretical information rather than brief experimental detail.

Cambridge IGCSE Chemistry Study and Revision Guide © David Besser

1 The particulate nature of matter

Key objectives

By the end of this section, you should

- know the different properties of solids, liquids and gases
- be able to describe the structure of solids, liquids and gases in terms of particle separation, arrangement and types of motion
- know what is meant by melting, boiling, evaporation, freezing, condensation and sublimation
- be able to describe the effect of temperature on the motion of gas particles

- have an understanding of Brownian motion
- be able to describe and explain diffusion
- be able to explain changes of state in terms of the kinetic theory
- be able to describe and explain Brownian motion in terms of random molecular bombardment and state evidence for Brownian motion
- be able to describe and explain dependence of rate of diffusion on relative molecular mass.

● Key terms

Melting	The process that occurs when a solid turns into a liquid
Melting point	The temperature at which a substance melts. Each substance has a specific melting point
Boiling	The process that occurs when a liquid turns into a gas
Boiling point	The temperature at which a substance boils. Each substance has a specific boiling point
Evaporation	The process that occurs at the surface of a liquid as it turns into a gas. Evaporation can occur at temperatures lower than the boiling point of a liquid
Freezing	The process that occurs when a liquid turns into a solid
Freezing point	The temperature at which a substance freezes. This has the same value as the melting point
Condensation	The process that occurs when a gas turns into a liquid
Sublimation	The process that occurs when a solid turns into a gas without first turning into a liquid
Brownian motion	The random motion of visible particles caused by bombardment by much smaller particles
Diffusion	The process that occurs when particles move from a region of high concentration to a region of low concentration

● Solids, liquids and gases

Differences between solids, liquids and gases are shown in Figure 1.1.

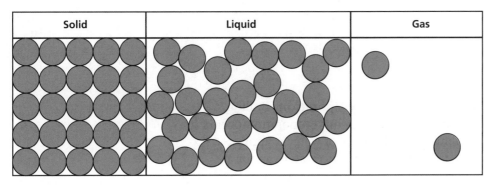

Figure 1.1 (a) Solid, (b) liquid, (c) gas

Cambridge IGCSE Chemistry Study and Revision Guide © David Besser

Examiner's tip

When asked to draw diagrams of the arrangement of particles in solids, liquids and gases, solids and gases are usually drawn quite well, but the particles in liquids are usually drawn too far apart. In reality, the majority of particles in a liquid are touching.

The differences between the properties of solids, liquids and gases, along with the reasons (based on kinetic theory) for the differences, are shown in Table 1.1.

Table 1.1 Properties of solids, liquids and gases

		Surface boundary	Shape
Solid	Property	Solids have a surface boundary	Solids have a fixed shape
	Reason	Strong forces of attraction between particles prevent particles from escaping	Strong forces of attraction between particles in solids mean that the particles are held together in a fixed shape. The particles vibrate about fixed positions but do not move from place to place
Liquid	Property	Liquids have a surface boundary	Liquids take the shape of the container that they are present in
	Reason	The forces of attraction between the particles in a liquid are strong enough to prevent the majority of the liquid particles from escaping and becoming a gas	The forces of attraction between particles in a liquid are weaker compared to solids. Therefore the particles slowly move from place to place meaning that a liquid can change its shape to fit the container
Gas	Property	Gases have no surface boundary	Gases fill the container they are held in. They have no fixed shape
	Reason	Gas particles move at high speeds. The particles have only very small forces of attraction between them	The forces of attraction between gas particles are extremely weak. The gas particles move at very high speeds therefore gases move to fill the container

● Changes of state

Figure 1.2 summarises the changes in state that occur between solids, liquids and gases.

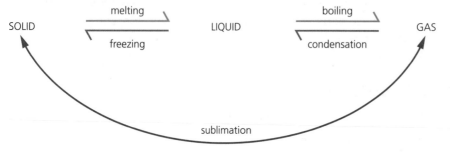

Figure 1.2 Changes of state

Common error

- There is often confusion between **boiling** and **evaporation**. Boiling only takes place at the **boiling point** of a liquid, but evaporation occurs at temperatures below the boiling point. Puddles of water evaporate on a sunny day. This means that the water turns into water vapour at temperatures well below the boiling point of water. The water in the puddles does not reach 100 °C!

● Heating and cooling curves

A **heating curve** shows the changes of state occurring when the temperature of ice is gradually increased. A similar (but not the same) curve results when a gas is cooled gradually until it forms a solid. This is known as a **cooling curve**.

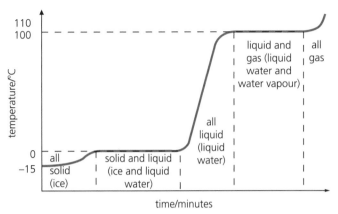

Figure 1.3 Graph of temperature against time for the change from ice at –15 °C to water to steam

The process begins with ice at a temperature below 0 °C. The temperature gradually increases until it reaches 0 °C, which is the **melting point** of ice. At this point ice and water exist together. The temperature does not change until all the ice has changed into water which is why the line is horizontal. A sharp melting point (at one specific temperature) is an indication that any solid is pure.

The temperature then begins to increase again until it reaches 100 °C which is the boiling point of water. The temperature does not change until all the water has changed into water vapour which is why the line is horizontal for a second time. When all the water has boiled, the temperature begins to rise again as the particles in the gaseous state gain more energy.

● Kinetic theory

When heat energy is given to a solid, the heat energy causes the particles to vibrate faster and faster about a fixed position until the particles have sufficient energy for **melting** to occur. At the melting point the energy gained by the particles is sufficient to overcome the attraction between particles in the solid. The ordered arrangement of particles then breaks down as the solid turns into a liquid. As this is occurring, there is no further increase in temperature until the ordered arrangement has completely broken down and all the solid has turned into a liquid. The energy given to the particles then causes them to move faster from place to place until they have sufficient energy for **boiling** to occur. At the boiling point the energy gained by the particles is sufficient to completely overcome the attraction between them in the liquid state. The particles then move as far away from each other as possible as the forces of attraction between them are almost completely overcome. Again there is no increase in temperature until the liquid has turned completely into a gas. In the gaseous state, the gas particles gain more and more energy and move at increasing speeds.

Cambridge IGCSE Chemistry Study and Revision Guide © David Besser

● Brownian motion

When Robert Brown used a microscope to observe pollen grains on the surface of water in 1827, he noticed that the pollen grains moved in a random manner. This random movement is known as **Brownian motion**. The same thing can be observed if smoke particles in air are observed through a microscope.

Brownian motion is caused by the larger particles (pollen grains or smoke particles) being bombarded by smaller particles (water molecules or air molecules). The smaller particles move in straight lines until they collide with the larger particles. Because more of the smaller particles may collide on one side of the larger particles than the other, the movement of the larger particles is random and unpredictable.

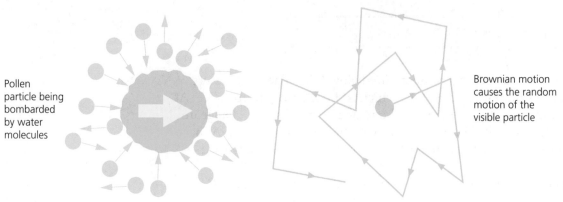

Pollen particle being bombarded by water molecules

Brownian motion causes the random motion of the visible particle

Figure 1.4 Brownian motion

● Diffusion

Particles in solids do not move from one place to another. However, particles in liquids move slowly and particles in gases move much faster.

Movement of particles from a region of high concentration to a region of low concentration is known as **diffusion**. It can be demonstrated experimentally in liquids and in gases.

Diffusion in liquids

If crystals of a coloured solid, such as nickel(II) sulfate, are placed in a liquid such as water, the colour of the nickel(II) sulfate spreads throughout the liquid in a matter of days, producing a solution with a uniform green colour.

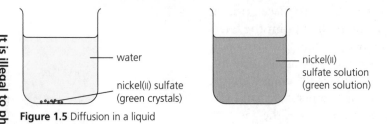

water

nickel(II) sulfate (green crystals)

nickel(II) sulfate solution (green solution)

Figure 1.5 Diffusion in a liquid

Diffusion in gases

If bromine liquid is placed in the bottom of a gas jar with another gas jar on top, the liquid evaporates and the brown colour of bromine gas fills both gas jars after a short time.

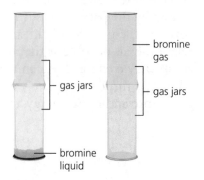

bromine gas

gas jars

gas jars

bromine liquid

Figure 1.6 Diffusion in gases

Cambridge IGCSE Chemistry Study and Revision Guide © David Besser

When gaseous molecules diffuse, the rate at which they diffuse is inversely related to the relative molecular mass of the gas. Therefore, molecules in gases with lower relative molecular mass will diffuse faster than molecules in gases with higher relative molecular mass. This is because lighter molecules move faster than heavier molecules.

● Sample exam-style questions

1 A compound has a melting point of –30 °C and a boiling point of 85 °C. What is its physical state at 25 °C? Explain your answer.

Student's answer

1 Liquid.
  The melting point is below 25 °C and the boiling point is above 25 °C.

Examiner's comments

There are two common errors in a question of this type.

● Some students ignore the negative sign in –30 °C, which gives them the impression that –30 °C is higher than 25 °C, which means they think the compound is a solid.
● Some students know that the substance is a liquid, but only state that the melting point is below 25 °C without mentioning the boiling point. Such candidates get some credit but not maximum credit.

2 When the apparatus shown in Figure 1.7 is set up, concentrated ammonia releases ammonia gas, NH_3, and concentrated hydrochloric acid releases hydrogen chloride gas, HCl.

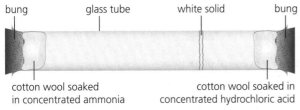

bung   glass tube   white solid   bung

cotton wool soaked in concentrated ammonia   cotton wool soaked in concentrated hydrochloric acid

Figure 1.7

When ammonia gas reacts with hydrogen chloride gas, a white solid is produced according to the equation:

$$NH_{3(g)} + HCl_{(g)} \rightarrow NH_4Cl_{(s)}$$

a What is the name of the white solid?

b Name the process by which the two gases move through the glass tube.

c Explain why the white solid forms nearer the concentrated hydrochloric acid end of the glass tube rather than the ammonia solution end.

Student's answer

2 a Ammonium chloride
 b Diffusion
 c Molecules of ammonia and hydrogen chloride diffuse through the glass tube. Because ammonia, NH_3, has a lower relative molecular mass (17) than hydrogen chloride, HCl (36.5), ammonia molecules diffuse faster than hydrogen chloride molecules. Therefore, the gases meet and react nearer the hydrochloric acid end.

Examiner's comments

a Ammonium compounds are often mistakenly referred to as *ammonia compounds*. Similarly ammonia is often referred to as *ammonium*. Students should make sure that they are aware of the difference between ammonia, NH_3, and the ammonium ion, NH_4^+, which is part of all ammonium salts such as ammonium chloride.

b This is the correct answer.

c It is common to see statements about ammonia moving faster than hydrogen chloride because ammonia is lighter than hydrogen chloride. This would gain very little (if any) credit. Answers must refer to ammonia and hydrogen chloride molecules and also state that ammonia has a smaller relative molecular mass than hydrogen chloride which is why ammonia molecules diffuse faster.

Students should calculate relative molecular masses, using relative atomic masses in the Periodic Table, if they are not provided in the question.

Exam-style questions

1 A substance has a melting point of 85 °C and a boiling point of 180 °C. What is the physical state of the substance at 50 °C? Explain your answer.   [Total: 2 marks]

2 Use the letters A, B, C and D to answer the questions under the table.

Substance	Distance between particles	Arrangement of particles	Movement of particles
A	very far apart	ordered	vibrate about fixed position
B	fairly close together	irregular	move slowly
C	very close together	ordered	vibrate about fixed position
D	very far apart	random	move at high speeds

Which substance out of A, B, C and D is

a a solid   [1]

b a liquid   [1]

c a gas   [1]

d unlikely to represent a real substance?   [1]
                                                                                          [Total: 4 marks]

Elements, compounds and experimental techniques

● Key terms

Element	A substance that cannot be decomposed into anything simpler by chemical means. It is a substance made up of atoms all of which have the same atomic number (see also Chapter 3)
Compound	A substance which contains two or more elements chemically combined in fixed proportions by mass
Mixture	Contains two or more substances (elements or compounds) which can be present in variable proportions
Substance	A general term that refers to elements, mixtures and compounds
Solution	A liquid which contains a substance or substances dissolved in it
Solvent	A pure liquid
Solute	The dissolved substance in a solution

● Elements

The periodic table consists of **elements** only. Each element has a chemical symbol. Elements are classified as metals and non-metals as shown in Table 2.1 below.

Table 2.1 Classification of elements

Property	Metal	Non-metal
Physical state at room temperature	Solid (except mercury)	Solid, liquid (bromine only) or gas
Malleability	Good	Poor, usually soft or brittle
Ductility	Good	Poor, usually soft or brittle
Appearance	Shiny (lustrous)	Usually dull
Melting point/boiling point	Usually high	Usually low
Density	Usually high	Usually low
Conductivity (electrical and thermal)	Good	Poor (except graphite)

Examiner's tip

Elements cannot be decomposed into anything simpler by chemical means. Students sometimes use the word smaller instead of simpler, which is an error. For example, a piece of sulfur can be broken with a hammer into several smaller pieces of sulfur, but this is not breaking it into anything simpler. The act of breaking with a hammer is a physical process and not a chemical process. Thus sulfur is an element.

● Compounds

Compounds have a chemical formula which shows them to contain two or more elements which are chemically combined.

Examples of compounds are:

- sodium chloride, NaCl
- carbon dioxide, CO_2
- copper(II) nitrate, $Cu(NO_3)_2$.

A compound, iron(II) sulfide, has the formula FeS. The **relative atomic masses** (A_r) of iron (Fe) and sulfur (S) are 56 and 32, respectively. Therefore, the **relative molecular mass** (M_r) of iron(II) sulfide is $56 + 32 = 88$. The percentages of iron and sulfur in iron(II) sulfide are:

$$Fe = (56 \div 88) \times 100 = 63.6\%$$
$$S = (32 \div 88) \times 100 = 36.4\%$$

This means that all samples of iron(II) sulfide contain 63.6% iron and 36.4% sulfur by mass.

This is what is meant by the statement that compounds contain elements chemically combined in fixed proportions by mass.

> **Examiner's tip**
>
> Students may state that compounds contain two or more elements, but often do not mention that the elements are chemically combined (which means that the elements are joined by ionic or covalent bonds).

> **Examiner's tip**
>
> If a compound is present in an aqueous **solution**, the aqueous solution is a **mixture** because it contains two **substances** which are not chemically combined. For example, sodium hydroxide solution (also referred to as aqueous sodium hydroxide) is a mixture, not a compound. It contains sodium hydroxide and water, two substances that are not chemically combined.

● Mixtures

Mixtures contain two or more elements and/or compounds in variable proportions. Mixtures do not have a chemical formula.

Air is an example of a mixture. Air contains nitrogen and oxygen with smaller amounts of other gases, such as water vapour, carbon dioxide and argon. Polluted air may also contain other gases such as carbon monoxide, sulfur dioxide and nitrogen dioxide.

Air has different percentages of its constituent gases in different places due to, for example, the amounts of pollutant gases which are lower in the countryside than in industrial areas. However, the different samples are all called 'air', thus showing that the composition of a mixture can vary.

Air does not have a chemical formula because it contains several chemical substances as opposed to one substance.

> **Examiner's tip**
>
> Many students are under the impression that a mixture containing two substances, such as salt and sand, must contain equal amounts of each substance. This is not the case. If we had a mixture of salt and sand which contained equal amounts of each substance and we added more salt to it, it would still be called a mixture of salt and sand. Therefore, a mixture of salt and sand can contain more salt than sand, or more sand than salt or equal amounts of salt and sand. This is different to the composition of a compound as shown in the case of iron(II) sulfide above.

● Separation of mixtures

Chromatography

Chromatography can be used to separate the components of solutions which contain several dissolved substances. The substances are often coloured, but may be colourless.

Paper chromatography can be used to separate the dyes in ink.

- A spot of the ink is placed on the chromatography paper.
- The paper is placed in a suitable solvent in a beaker. If the solvent is volatile (vaporises easily) it is necessary to put a lid on the beaker to prevent the vapour from escaping.
- As the solvent rises, the dyes in the ink separate.

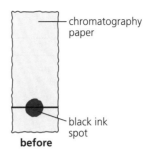

Chromatography can also be used to identify the components of a mixture as well as separate them.

- A mixture of dyes is placed on chromatography paper in the position marked X.
- Four dyes whose identities are known are placed in positions marked A, B, C and D, as shown (Figure 2.2). These four dyes are referred to as standards.
- Chromatography is then carried out and the chromatography paper (also known as a chromatogram) is removed from the beaker and dried.
- The paper is then labelled to show what mixture X contains, as described below.

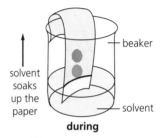

before

during

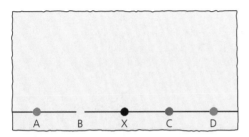

Figure 2.2 Before chromatography

after

Figure 2.1 Paper chromatography

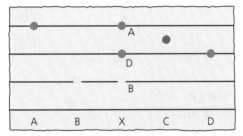

Figure 2.3 After chromatography

The results of the experiment show:

- X is composed of three dyes because the mixture has been separated into three.
- The three dyes are A, B and D. We know this because the three dyes in mixture X have travelled the same distances as the three standards A, B and D whose identities are known.
- We can also conclude that mixture X does not contain dye C, because none of the components of X travelled the same distance as dye C.

Chromatography can also be used to identify colourless substances. The experimental technique is the same, but because the components of the mixture are colourless, the spots on the chromatography paper are invisible. After drying, the paper is sprayed with a **locating agent** which reacts with the components of the mixture to produce coloured spots. In Chapter 15 which looks at the separation of amino acids by chromatography, the chromatography paper is sprayed with ninhydrin which is a locating agent that produces blue coloured spots with amino acids.

Instead of using standards as described in the above experiment, components of a mixture can be identified by their R_f **values**. Chromatography is carried out and after the chromatography paper is dried, the distance that the solvent has travelled and the distance that the component of the mixture has travelled are both measured as shown in Figure 2.4.

Cambridge IGCSE Chemistry Study and Revision Guide © David Besser

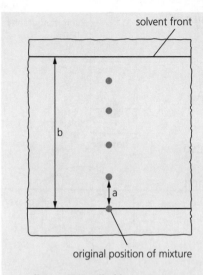

solvent front

b

a

original position of mixture

$$R_f = \frac{\text{distance travelled by component}}{\text{distance travelled by solvent}} = \frac{a}{b}$$

Figure 2.4 Calculating R_f

Examiner's tip

When asked what is meant by R_f value, students are advised to write down the equation given here, rather than trying to explain R_f value in the form of a sentence, which is much more difficult. The correct equation would score all the available marks.

This illustrates the fact that answers need not always be expressed in words, sentences and paragraphs alone to get full credit. Diagrams, equations, sketch graphs and formulae are often much more appropriate than sentences.

When the R_f value is calculated, the component of the mixture can be identified by comparison with R_f values in a data book. R_f values can be determined for all the components of the mixture.

Dissolving, filtration and crystallisation

Dissolving, **filtration** and **crystallisation** are methods used to separate a mixture of two solids, one of which is soluble in a given solvent and the other of which is insoluble.

This method can be used to separate a mixture of common salt and sand and produce pure samples of both solids.

- If the mixture is not powdered it should be ground into a powder using a mortar and pestle. The powder should be added to water in a beaker. The common salt dissolves and the sand remains undissolved.
- The mixture is then transferred to the filtration apparatus. The sand (**residue**) remains in the filter paper and the salt solution (**filtrate**) passes through into the conical flask. This process is called filtration.
- To obtain pure sand, distilled water should be passed through the filter paper (this is known as washing the residue) and then the filter paper should be removed and dried in a low oven or on a warm windowsill.

The final two stages are known as crystallisation:

- To obtain salt crystals, the salt solution should then be heated in an evaporating dish until about half of the water has been removed (alternatively, when crystals form on a glass rod placed in the hot solution and withdrawn, it is time to stop heating).
- The hot saturated salt solution should then be allowed to cool down slowly. Crystals of salt should then form.
- If there is any liquid left, it should then be separated by filtration. The salt crystals should then be washed with a small amount of cold distilled water and then dried in a low oven or on a warm windowsill.

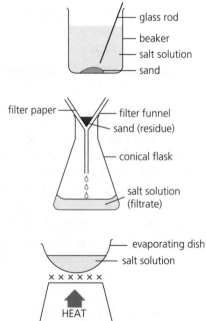

glass rod
beaker
salt solution
sand

filter paper
filter funnel
sand (residue)

conical flask

salt solution
(filtrate)

evaporating dish
salt solution

HEAT

Figure 2.5 Dissolving, filtration and crystallisation

Cambridge IGCSE Chemistry Study and Revision Guide © David Besser

Common errors

Filtration (often spelt wrongly as filteration) can also be called filtering. There are other common errors when describing the process:

- The words *residue* and *filtrate* are often used the wrong way round.
- *Filtrate* is often used as an incorrect alternative to *filtered* as in 'he filtrated the solution'. The word filtrated does not exist!
- The filtrate should not be heated until all the water evaporates to dryness. This does not lead to the production of good crystals. In addition, some crystals contain water of crystallisation which would be driven off by too much heat (see Chapter 8).
- If the crystals are dried with too much water they will dissolve, which defeats the purpose. The water should ideally be ice cold to minimise the amount that dissolves.

(Simple) distillation

(Simple) distillation is a method of separating a pure liquid from a solution.

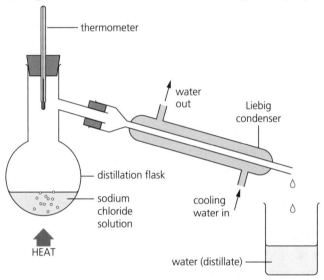

Figure 2.6 Simple distillation

The flask is heated. The water in the sodium chloride solution evaporates and water vapour/steam enters the Liebig condenser, where it condenses as water. The water drips out of the end of the Liebig condenser and collects in the beaker. The water is pure and can be called **distilled water**. Sodium chloride does not vaporise or even melt because it has a very high melting point, and therefore it remains in the distillation flask.

Fractional distillation

Fractional distillation is a method of separating two (or more) miscible liquids with different boiling points. It can be carried out in the laboratory or on an industrial scale as in the fractional distillation of liquid air (see Chapter 11) or fractional distillation of petroleum (see Chapter 14).

In the laboratory ethanol and water can be separated by fractional distillation using the apparatus shown in Figure 2.7.

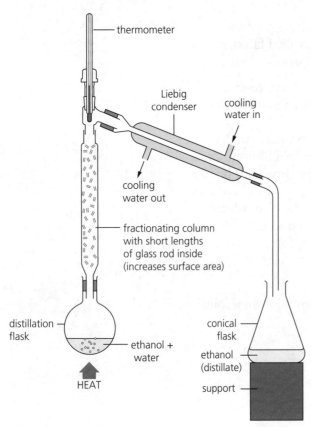

Figure 2.7 Apparatus for fractional distillation

Ethanol has a boiling point of 78 °C and water has a boiling point of 100 °C.
The flask is heated and ethanol vapour enters the fractionating column.
However, some water also evaporates (below its boiling point) and enters
the fractionating column as water vapour/steam. The water vapour/
steam condenses in the fractionating column and drips back down into the
distillation flask. When the temperature reaches 78 °C, the ethanol vapour
reaches the top of the fractionating column and enters the Liebig condenser
where it condenses. Finally, liquid ethanol collects as the distillate and all the
water remains in the distillation flask.

● Summary: Methods of separation of mixtures

Method of separation	Example of mixture that is separated	Property that the method depends on
Paper chromatography	Dyes in ink	Adsorption by paper/solubility in solvent
Dissolving, filtration and crystallisation	Sand and salt	Solubility
(Simple) distillation	Sodium chloride solution	Boiling point
Fractional distillation	Ethanol and water	Boiling point

● Sample exam-style question

1 State whether the following are elements, mixtures or compounds.

a silver

b bronze

c sea water

d water

e bauxite

f aluminium oxide

Student's answer

1 a element
   b mixture
   c mixture
   d compound
   e mixture
   f compound

Examiner's comment

a Metals can be elements, but alloys are mixtures of metals. Silver is an element. It is the element with atomic number 47 in the Periodic Table. If you are not sure if a substance is an element you should know that the Periodic Table only contains elements.

b Bronze is an alloy (see Chapter 10) and, as such, it is a mixture of metals. Bronze contains copper, tin and other metals in variable proportions.

c Sea water is water (which is a compound) containing many substances, in variable proportions, dissolved in it.

d Water has the formula H_2O. Any substance with a formula that shows more than one element is a compound. Although water is found in many forms such as tap water, sea water and distilled water, the term water refers to the pure compound.

e Bauxite is a metallic ore from which aluminium is extracted (see Chapter 5). The word ore refers to an impure substance. Metallic ores are mixtures.

f Aluminium oxide has the formula Al_2O_3. Bauxite contains the compound aluminium oxide with impurities.

Exam-style questions

1 State the name of the process(es) that you would use to obtain

   a sugar crystals from a mixture of sugar and sand

   b pure water from an aqueous solution of copper(ɪɪ) sulfate

   c liquid octane (boiling point 126 °C) from a mixture of liquid octane and liquid decane (boiling point 174 °C)

   d pure silver chloride from the precipitate formed when aqueous silver nitrate is added to dilute hydrochloric acid.

   In some cases, only one process is required, but others may require more than one. [Total: 8 marks]

2 A student was told to make pure crystals of copper(ɪɪ) sulfate from an aqueous solution of copper(ɪɪ) sulfate. Describe how the student should carry this out. [4 marks]

3 A student is given a mixture of two amino acids. The amino acids are both colourless solids that are soluble in water. Give full experimental details of how you would separate and identify the amino acids present in the mixture using paper chromatography. You are provided with all the necessary apparatus and a suitable locating agent. [5 marks]

Cambridge IGCSE Chemistry Study and Revision Guide © David Besser

Atomic structure, bonding and structure of solids

Key objectives

By the end of this section, you should

- know the relative charges and approximate relative masses of protons, neutrons and electrons
- know the definition of proton number (atomic number) and nucleon number (mass number)
- know that proton number is the basis of the Periodic Table
- know the definition of isotopes and that isotopes are radioactive and non-radioactive
- know one medical and one industrial use of radioactive isotopes
- know the build up of electrons in shells for the first 20 elements in the Periodic Table
- know the difference between metals and non-metals
- know that ions are formed when atoms lose and gain electrons
- be able to describe the formation of ionic bonds between elements from Groups I and VII
- be able to describe the formation of single covalent bonds in H_2, Cl_2, H_2O, CH_4, NH_3, HCl
- be able to describe the difference in volatility, solubility and electrical conductivity between ionic compounds and covalent substances with both giant structures and simple molecular structures
- know the giant covalent structures of graphite and diamond

- relate the structures of graphite and diamond to their uses
- know why isotopes have the same chemical properties
- be able to describe the formation of ionic bonds between other metallic and non-metallic elements
- be able to determine the formulae of ionic compounds from the charges on the ions present
- know that a giant ionic lattice is a regular arrangement of positive and negative ions
- be able to describe the formation of more complex covalent molecules, such as N_2, C_2H_4, CH_3OH and CO_2
- be able to explain the differences in melting point and boiling point between ionic compounds and covalent substances with both giant structures and simple molecular structures in terms of attractive forces between particles
- know the structure of silicon(IV) oxide (silicon dioxide)
- be able to describe the similarity in properties between diamond and silicon(IV) oxide (silicon dioxide) related to their structures
- be able to describe metallic bonding as a force of attraction between positive ions and a mobile sea of electrons
- know how the structure of metals can be used to explain malleability and conduction of electricity.

Key terms

Proton number (atomic number)	The number of protons in one atom of an element
Nucleon number (mass number)	The sum of the number of protons and neutrons in one atom of an element
Isotopes	Atoms of the same element containing the same number of protons but different numbers of neutrons, or Atoms of the same element with the same proton number (atomic number) but different nucleon number (mass number)
Lattice	A regular arrangement of particles present in a solid. The particles (atoms, molecules or ions) are arranged in a repeated pattern

Atomic structure

Atoms are made from smaller particles called **protons, neutrons** and **electrons**.

Table 3.1 The properties of protons, neutrons and electrons

Particle	Relative mass/atomic mass units	Relative charge
Proton	1	+1
Neutron	1	0
Electron	1/1837 (negligible)	−1

Examiner's tip

Make sure you learn the information in Table 3.1. You need to know the differences between **relative mass** and **relative charge** of a proton, neutron and electron.

Cambridge IGCSE Chemistry Study and Revision Guide © David Besser

The protons and neutrons exist in the centre of the atom in a dense region called the **nucleus**. The electrons move around the nucleus and exist in **electron shells** at increasing distances from the nucleus.

Atoms are often represented as shown in Figure 3.1.

nucleon number (mass number)

proton number (atomic number)

Figure 3.1

Examiner's tip

In some books, the two numbers may be reversed. It is a good idea to remember that the **nucleon number** is always higher than the **proton number** (with the exception of hydrogen, in which case both numbers are 1 in the most abundant isotope).

The proton number is the number of protons in one atom of the element. Because atoms do not have a charge, the number of protons in an atom is always equal to the number of electrons.

The nucleon number is the number of neutrons and protons added together in one atom of an element.

Therefore

Proton number = number of protons in one atom = number of electrons in one atom

Number of neutrons = nucleon number – proton number

In the example given in Figure 3.1:

Number of protons = proton number	Number of electrons = number of protons	Number of neutrons = (nucleon number – proton number)
15	15	31 – 15 = 16

Isotopes are atoms of the same element containing the same number of protons but different numbers of neutrons.

Examples of isotopes of argon are shown in Table 3.2.

Table 3.2 Examples of isotopes of argon

Isotope	Number of protons in one atom	Number of neutrons in one atom	Number of electrons in one atom
$^{40}_{18}$Ar	18	(40 – 18) = 22	18
$^{38}_{18}$Ar	18	(38 – 18) = 20	18
$^{36}_{18}$Ar	18	(36 – 18) = 18	18

Some isotopes are radioactive and some are non-radioactive. Radioactive isotopes decay (which means give off radiation) and usually change into other elements. Radioactive isotopes are also known as **radioisotopes**.

Radioactive isotopes/radioisotopes can be used

- in medicine (e.g. cobalt-60 is used in radiotherapy treatment)
- in industry (e.g. uranium-235 is used as a source of power in nuclear reactors).

Common error

- Many students think of isotopes as referring only to radioactive isotopes, but some are also non-radioactive.

● The arrangement of electrons in atoms

Electrons are arranged in electron shells at increasing distances from the nucleus. These shells can hold up to a maximum number of electrons, as shown in Table 3.3.

Cambridge IGCSE Chemistry Study and Revision Guide © David Besser

Table 3.3 Maximum number of electrons per shell number

Shell number	Maximum number of electrons
1	2
2	8
3	8*

*Shell 3 can, in fact, hold up to 18 electrons, but this does not need to be considered at this level and only becomes relevant in higher-level courses.

Some examples of arrangement of electrons in shells are shown Table 3.4.

Table 3.4 Arrangement of electrons in shells

Element	Number of electrons in one atom	Arrangement of electrons in shells
Helium, He	2	2
Carbon, C	6	2,4
Phosphorus, P	15	2,8,5
Potassium, K	19	2,8,8,1

Students are expected to be able to write down and draw the electron arrangement of the first 20 elements in the Periodic Table.

Isotopes of the same element all have the same number of electrons, and therefore all have the same number of electrons in their outer shells. This means that isotopes of the same element all have the same chemical properties.

The chemical properties of elements depend on the number of electrons in the outer shell of their atoms.

● The Periodic Table

Elements in the Periodic Table are arranged in order of increasing proton number. This means that as we move from one element to the next element, the atoms have one extra proton in the nucleus and one extra electron. The extra electron goes into the outer shell until the outer shell is full. The next shell then begins to fill up.

Elements in the same group all have the same number of electrons in the outer shell of their atoms. This applies beyond the first 20 elements. Examples are:

● All Group (I) elements have 1 electron in the outer shell.
● All Group (II) elements have 2 electrons in the outer shell.
● All Group (VII) elements have 7 electrons in the outer shell.
● All Group (0) elements have a full outer shell.
● Most metallic elements have 1, 2 or 3 electrons in their outer shell.
● Most non-metallic elements have 5, 6 or 7 electrons in their outer shell
  (or a full outer shell in the case of Noble gases).

● Ionic bonding

Ionic bonding occurs in compounds containing metallic elements combined with non-metallic elements.

Metal atoms (with 1, 2 or 3 electrons in their outer shells) lose an electron or electrons in order to achieve a full outer shell and form positive ions (**cations**).

Non-metal atoms (with 5, 6 or 7 electrons in their outer shells) gain an electron or electrons in order to achieve a full outer shell and form negative ions (**anions**). An example occurs in sodium chloride.

Sodium atoms contain 11 protons and 11 electrons and chlorine atoms contain 17 protons and 17 electrons. Because both contain equal numbers of protons and electrons both atoms are uncharged.

An electron moves from the outer shell of a sodium atom to the outer shell of a chlorine atom so that both atoms achieve a full outer shell.

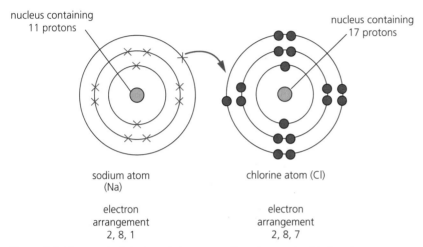

Figure 3.2 Movement of electrons between a sodium atom and a chlorine atom

After the transfer of electrons, sodium forms a positive sodium ion and chlorine forms a negative chlor**ide** ion.

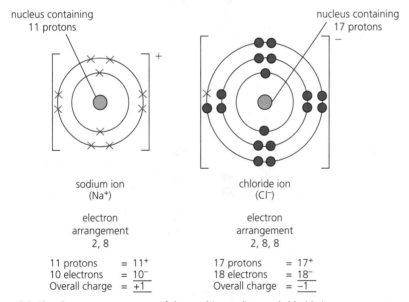

11 protons	= 11⁺
10 electrons	= 10⁻
Overall charge	= +1

17 protons	= 17⁺
18 electrons	= 18⁻
Overall charge	= −1

Figure 3.3 The electron arrangements of the resulting sodium and chloride ions

The sodium ion still has 11 protons but only 10 electrons, therefore it has 1+ charge and is written Na^+. The chlor**ide** ion still has 17 protons but now has 18 electrons; therefore it has 1⁻ charge and is written Cl^-.

In sodium chloride, the ratio of sodium ions to chloride ions is 1:1 and thus the formula of sodium chloride is NaCl. In all examples in which Group (I) elements combine with Group (VII) elements, the ratio of ions is always 1:1.

Other examples occur in which atoms do not combine in the ratio 1:1. This applies when the number of electrons lost by one metal atom is not equal to the number of electrons gained by one non-metallic atom.

Examiner's tip

Remember that atoms have equal numbers of protons and electrons, and are therefore uncharged.

Positive ions (cations) have more protons than electrons and are therefore positively charged.

Negative ions (anions) have more electrons than protons and are therefore negatively charged.

Cambridge IGCSE Chemistry Study and Revision Guide © David Besser

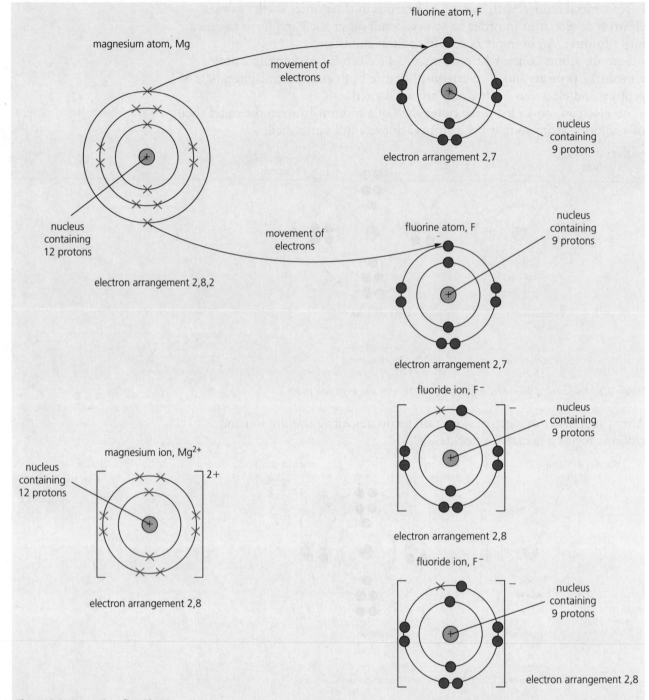

Figure 3.4 Magnesium fluoride

Because the ratio of magnesium ions to fluoride ions is 1:2, the formula of magnesium fluoride is MgF_2.

The formulae of ionic compounds

The formulae of ionic compounds can be deduced from knowledge of the charges on the ions. Examples of common ions are shown in Table 3.5.

Cambridge IGCSE Chemistry Study and Revision Guide © David Besser

Table 3.5 Examples of common ions

Charge	Example	Charge	Example
1+	Lithium, Li^+ Sodium, Na^+ Potassium, K^+ Silver, Ag^+ Ammonium, $\mathbf{NH_4^+}$	1–	Fluoride, F^- Chloride, Cl^- Bromide, Br^- Iodide, I^- Hydroxide, $\mathbf{OH^-}$ Nitrite, $\mathbf{NO_2^-}$ Nitrate, $\mathbf{NO_3^-}$
2+	Magnesium, Mg^{2+} Calcium, Ca^{2+} Barium, Ba^{2+} Zinc, Zn^{2+} Iron(ıı), Fe^{2+} Copper(ıı), Cu^{2+} Lead, Pb^{2+}	2–	Oxide, O^{2-} Sulfide, S^{2-} Carbonate, $\mathbf{CO_3^{2-}}$ Sulfate, $\mathbf{SO_4^{2-}}$ Sulfite, $\mathbf{SO_3^{2-}}$
3+	Aluminium, Al^{3+} Iron(ııı), Fe^{3+}	3–	Nitride, N^{3-} Phosphate, $\mathbf{PO_4^{3-}}$

Bold type denotes polyatomic ions. These are ions which have more than one capital letter in the formula (see rule 4 below).

Examples of how to determine formulae of ionic compounds

The most important thing to know is that all compounds have no overall charge; therefore in the case of ionic compounds the number of positive charges is equal to the number of negative charges.

To work out the formula of a compound you should:

1 Write down the formulae of the positive and the negative ions.

2 Count the number of positive charges and the number of negative charges.

3 If the charges are not equal, add more positive ions, more negative ions or both until the charges are equal.

4 If more than one of a polyatomic ion is required, the whole formula of the ion must go in a bracket and the number of ions goes outside the bracket as a subscript, e.g. $(NO_3)_2$.

Write down the formulae of the following compounds.

Sodium carbonate				
	1	Ions	Na^+	CO_3^{2-}
	2	Charges	$Na\textcircled{+}$ ── 1+	$CO_3\textcircled{2-}$ ── 2–
	3	Add 1 extra Na^+ to make the charges equal	$Na\textcircled{+}$ $Na\textcircled{+}$ ── 2+	$CO_3\textcircled{2-}$ ── 2–
	4	There is only one (CO_3^{2-}) therefore a bracket is not required		
	5	Formula Na_2CO_3		

Examiner's tip

Table 3.6 shows the charges on ions in different groups in the Periodic Table.

Table 3.6

Group	Charge on ion
I	1+
II	2+
III	3+
V	3–
VI	2–
VII	1–

In other cases, e.g. the transition elements, it is not possible to use the Periodic Table to deduce the charges on ions. In such cases, the charges must be learned by heart.

Cambridge IGCSE Chemistry Study and Revision Guide © David Besser

Magnesium hydroxide	1	Mg^{2+}		OH^-
	2	$\dfrac{Mg^{2+}}{2+}$		$\dfrac{OH^{\ominus}}{1-}$
	3	Mg^{2+}		$OH^{\ominus}$
				$\dfrac{OH^{\ominus}}{2-}$
		$\overline{2+}$		
	4	OH^- has two capital letters. Since $2OH^-$ ions are required, OH must go in a bracket with the 2 outside the bracket as a subscript.		
	5	$Mg(OH)_2$		
Aluminium oxide	1	Al^{3+}		O^{2-}
	2	$\dfrac{Al^{3+}}{3+}$		$\dfrac{O^{2-}}{2-}$
	3	Al^{3+}		O^{2-}
		Al^{3+}		O^{2-}
				O^{2-}
		$\overline{6+}$		$\overline{6-}$
	4	There are no polyatomic ions, therefore brackets are not required.		
	5	Al_2O_3		
Iron(III) sulfate	1	Fe^{3+}		SO_4^{2-}
	2	$\dfrac{Fe^{3+}}{3+}$		$\dfrac{SO_4^{2-}}{2-}$
	3	Fe^{3+}		SO_4^{2-}
		Fe^{3+}		SO_4^{2-}
				SO_4^{2-}
		$\overline{6+}$		$\overline{6-}$
	4	Because (SO_4^{2-}) has two capital letters, we need to put SO_4 in a bracket with 3 outside the bracket as a subscript.		

Common errors

Some common incorrect answers using the formula of iron(III) sulfate as an example are

- $FeSO_4$: The number of charges has not been made equal. (This is the correct formula of iron(II) sulfate.)
- $Fe_2(SO)_3$: The 4 is left out of the formula of sulfate.
- $(Fe)_2(SO_4)_3$: A bracket is not required around Fe as it only has one capital letter.
- $Fe_2(SO)_4$: The 4 is left out of the formula of sulfate and placed incorrectly outside the bracket.

Examiner's tip

You should be able to write formulae of compounds containing all possible combinations of positive and negative ions in Table 3.5.

Cambridge IGCSE Chemistry Study and Revision Guide © David Besser

● Covalent bonding

Covalent bonding occurs in elements and compounds containing non-metallic elements only.

Covalent bonds are formed when pairs of electrons are shared. A shared pair of electrons is known as a single (covalent) bond.

Double bonds (two shared pairs of electrons) and triple bonds (three shared pairs of electrons) also exist.

Atoms which form a covalent bond join together to form uncharged molecules. All the atoms involved achieve a full outer shell of electrons.

Examples of covalent molecules containing single bonds only are shown in Figure 3.5. Only the outer electron shells are shown.

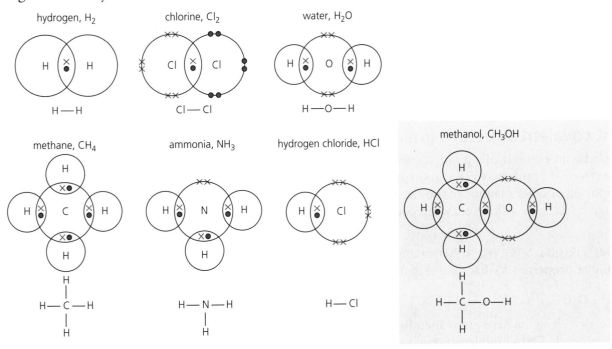

Figure 3.5 Covalent molecules containing single bonds

Examples of covalent molecules containing double and triple bonds are shown in Figure 3.6. Only the outer electron shells are shown.

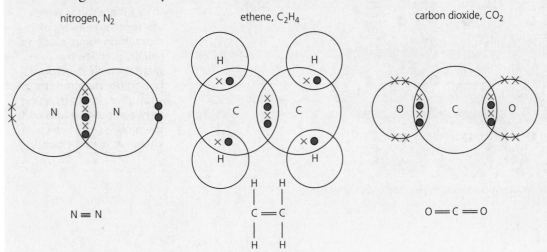

Figure 3.6 Covalent molecules containing double and triple bonds

Cambridge IGCSE Chemistry Study and Revision Guide © David Besser

● Structure of solids

Solids have four different types of structures as shown below.

Giant ionic structure

Sodium chloride is an example of a **giant ionic structure**. It is held together by strong forces of attraction between oppositely charged sodium ions and chloride ions (called ionic bonds) which are present in a giant ionic **lattice**.

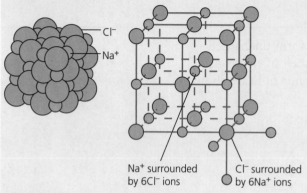

Na^+ surrounded by $6Cl^-$ ions

Cl^- surrounded by $6Na^+$ ions

Figure 3.7 The structure of sodium chloride

Giant covalent structure (macromolecular structure)

Diamond is an example of a **giant covalent structure**. It is held together by strong covalent bonds between carbon atoms.

Graphite is another example of substance with a giant covalent structure, although it has many differences when compared to diamond (see pages 23–24).

Silicon(IV) oxide, SiO_2, is another example of a giant covalent structure. It has similar properties to diamond due to a similar structure.

Giant metallic structures

All metallic elements have **giant metallic structures**. They contain positive ions surrounded by a mobile sea of electrons. Metals are held together by the strong forces of attraction between positive ions and the mobile sea of electrons known as **metallic bonds**.

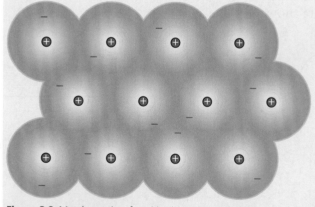

Figure 3.8 Metals consist of positive ions surrounded by a 'sea' of electrons

Examiner's tip

In exam questions which ask for the meaning of metallic bonding, students usually describe the giant metallic structure but do not mention the strong forces of attraction between positive ions and the mobile sea of electrons, known as metallic bonds.

Cambridge IGCSE Chemistry Study and Revision Guide © David Besser

Simple molecular structures

Iodine is an example of a substance with a **simple molecular structure**. It has strong covalent bonds between the atoms within the molecules (intramolecular) but weak intermolecular forces of attraction between the molecules.

The properties of different types of solid, related to their structures, are shown in Table 3.7.

Table 3.7 Properties of different types of solid, related to their structures

	Giant ionic structure	Giant molecular structure	Giant metallic structure	Simple molecular structure
Example	Sodium chloride, NaCl	Diamond, C Silicon dioxide, SiO_2	Copper, Cu	Iodine, I_2
Particles present	Positive and negative ions	Atoms	Positive ions and mobile sea of electrons	Simple molecules
Bonding	Strong forces of attraction between oppositely charged ions known as ionic bonds	Strong covalent bonds between atoms	Strong metallic bonds between positive ions and the mobile sea of electrons	Weak forces of attraction (intermolecular forces) between uncharged molecules
Solubility in water	Usually soluble	Insoluble	Insoluble. Note: some metals react with water	Usually insoluble
Melting point/ boiling point	High because all the bonds are strong ionic bonds	High because all the bonds are strong covalent bonds	High because all the bonds are strong metallic bonds	Low because intermolecular forces are weak and these are the bonds that break when these substances melt and boil, not the covalent bonds
Conduction of electricity	Only conduct when molten or dissolved in water, but not when solid (see Chapter 5)	Non-conductors (except graphite) because they do not contain mobile electrons or ions	Good conductors when solid because they contain mobile electrons (see Chapter 5)	Non-conductors because they do not contain mobile electrons or ions and consist of uncharged molecules
Malleability/ ductility	Not malleable or ductile	Not malleable or ductile	Malleable and ductile because rows of positive ions can slide over each other (see Chapter 10)	Not malleable or ductile

Common errors

- In exam questions which ask why substances with simple molecular structures have low melting points and boiling points, it is very commonly said that this is because covalent bonds are weak. This is a bad error. All covalent bonds are strong bonds.
- The correct answer is that intermolecular forces are weak which is why substances with simple molecular substances are either solids with low melting points such as iodine, liquids such as water or gases such as carbon dioxide.

● Diamond and graphite

Both diamond and graphite have giant covalent (macromolecular) structures, but because there are differences in their structure and bonding, these lead to differences in properties and uses.

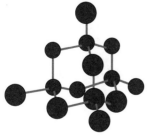

Figure 3.9 The structure of diamond

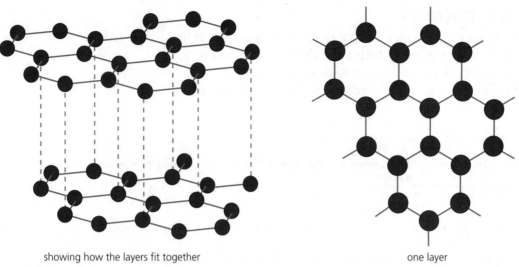

showing how the layers fit together   one layer

Figure 3.10 The structure of graphite

Table 3.8 Differences in structure and bonding between diamond and graphite

	Diamond	Graphite
Number of carbon atoms covalently bonded to each carbon atom	4	3
Arrangement of atoms	Tetrahedrally	In layers (made of rings containing 6 carbon atoms)
Bonding	All covalent	Covalent between atoms within layers Weak van der Waals forces between layers
Mobile electrons	None. All outer shell electrons used in bonding	One electron from each atom exists in the spaces in between the layers as mobile electrons
Hardness	Hard because all bonds are strong and directional	Soft because weak van der Waals forces between layers allow layers to slide over each other
Conduction of electricity	Non-conductor because there are no mobile electrons	Good conductors due to mobile electrons between layers
Use	In cutting tools due to high strength	As a lubricant because layers can slide As a conductor in motors

● Silicon(IV) oxide (silicon dioxide)

Silicon(IV) oxide (silicon dioxide) has a giant covalent (macromolecular) structure.

Each silicon atom is covalently bonded to four oxygen atoms. The bonds are directed tetrahedrally. Each oxygen atom is covalently bonded to two silicon atoms.

All the bonds in silicon(IV) oxide are strong covalent bonds. There are no mobile electrons present. Because of its structure and bonding, silicon(IV) oxide is strong, hard, has a high melting and boiling point and is a non-conductor of electricity. These properties are like those of diamond which has a very similar structure and the same bonding.

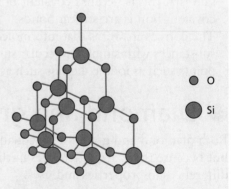

● O

● Si

Figure 3.11 The structure of silicon(IV) oxide

Cambridge IGCSE Chemistry Study and Revision Guide © David Besser

Exam-style questions

1 Complete the following table

Particle	Number of protons	Number of electrons	Electronic configuration	Charge on particle
A	20	18		
B	9	10	2,8	
C		10	2,8	0
D	8	10	2,8	

[Total: 5 marks]

2 Draw dot and cross diagrams showing the arrangement of outer shell electrons in the following covalent molecules.

a oxygen, O_2

d phosphine, PH_3

b methanol, CH_3OH H — C — O — H (with H above and H below the C)

e carbonyl chloride, $COCl_2$ (Cl above C, C=O, Cl below)   [Total: 5 marks]

c hydrogen cyanide, HCN   H—C≡N

3 Deduce the formulae of the following ionic compounds.

a magnesium hydroxide

b calcium chloride

c ammonium phosphate

d lithium sulfide

e lead nitrate

f calcium carbonate

g aluminium nitrate

h potassium sulfite

i zinc sulfate

j ammonium sulfate   [Total: 10 marks]

4 Use the table to answer the questions that follow. Consider room temperature to be 25 °C.

	Melting point/°C	Boiling point/°C	Conduction of electricity when solid	Conduction of electricity when molten
A	−40	35	Non-conductor	Non-conductor
B	50	150	Non-conductor	Non-conductor
C	801	1500	Non-conductor	Conductor
D	1500	2500	Conductor	Conductor
E	2500	5000	Non-conductor	Non-conductor
F	−75	−35	Non-conductor	Non-conductor

a Which substance or substances are solid at room temperature?   [1 mark]

b Which substance or substances are liquid at room temperature?   [1 mark]

c Which substance or substances are gaseous at room temperature?   [1 mark]

d Which substance could have a giant metallic structure?   [1 mark]

e Which substance has a giant ionic structure?   [1 mark]

f Which substance has a giant molecular structure?   [1 mark]

[Total: 6 marks]

Cambridge IGCSE Chemistry Study and Revision Guide © David Besser

Stoichiometry: chemical calculations

Key objectives

By the end of this section, you should

- be able to construct word equations and simple balanced equations
- be able to define relative atomic mass, A_r
- be able to define and calculate relative molecular mass, M_r
- be able to construct equations with state symbols, including ionic equations
- be able to deduce a balanced equation for a chemical reaction given relevant information
- be able to define the mole and the Avogadro constant and be able to use the molar gas volume (taken as $24\,dm^{-3}$ at room temperature and pressure)
- be able to calculate reacting masses, volumes of gases and solutions and concentrations of solutions
- define and calculate empirical formulae and molecular formulae
- be able to use the idea of limiting reactants
- calculate percentage yield and percentage purity.

● Key terms

Stoichiometry	The calculation of the relative quantities of reactants and products in a chemical reaction
Relative atomic mass	The average mass of one atom of an element on a scale where one ^{12}C atom has 12 units of mass exactly, A_r
Relative molecular mass	The sum of the relative atomic masses, M_r
Empirical formula	The smallest whole number ratio of the atoms of each element in a compound
Molecular formula	The number of atoms of each element in one molecule of a substance
Mole	The same number of particles as there are atoms in $12\,g$ of the carbon-12 isotope
Avogadro constant	The number of particles in one mole of a substance. It is equal to 6.02×10^{23} particles

● Stoichiometry

Stoichiometry means the calculation of the relative quantities of reactants and products in a chemical reaction.

Word equations

Word equations give the names of the reactants and products which take part in a chemical reaction. When hydrogen burns in oxygen to form water the word equation is:

hydrogen + oxygen → water

Symbol equations

Symbol equations give the correct formulae of the reactants and products in a reaction. Symbol equations are balanced when the number of atoms of each element is the same on both sides of the equation.

Steps for writing balanced equations

1 Write down the word equation (this can be omitted with experience).

2 Write down the correct formulae of reactants and products.

3 Count the number of atoms of each element on both sides.

4 If the number of atoms of each element on both sides is not the same, put numbers in front of the formulae so that the number of atoms of each element on both sides is the same.

5 Put state symbols after the formulae, i.e. (s) = solid, (l) = liquid, (g) = gas, (aq) = aqueous solution. This can be done after steps 2 or 3 if preferred.

Cambridge IGCSE Chemistry Study and Revision Guide © David Besser

● Worked Example

$$H_2 \quad + \quad O_2 \quad \rightarrow \quad H_2O$$

Number of atoms of each element on both sides	H 2 O 2	H 2 O 1

Because the number of atoms of oxygen is not the same on both sides, the first step is to put **2** in front of H_2O. This multiplies everything that comes after it.

$$H_2 \quad + \quad O_2 \quad \rightarrow \quad 2H_2O$$

Number of atoms of each element on both sides	H 2 O 2	H 4 O 2

In balancing the oxygen, we have unbalanced the hydrogen. Therefore we need to put a **2** in front of H_2. The equation is then balanced.
State symbols can be inserted.

$$2H_{2(g)} + \quad O_{2(g)} \quad \rightarrow \quad 2H_2O_{(l)}$$

Number of atoms of each element on both sides	H 4 O 2	H 4 O 2

Common errors

● Students often use incorrect formulae, e.g. H instead of H_2 or O instead of O_2, or change formulae such as changing H_2O into H_2O_2. This would make the number of atoms of each element the same on both sides, but H_2O_2 is not the correct formula for water (in fact, it is the formula for hydrogen peroxide). The only way to balance equations is to put numbers in front of the formulae.

● Worked Example

	aluminium	+	chlorine	→	aluminium chloride
Unbalanced:	Al	+	Cl_2	→	$AlCl_3$
	Al 1		Cl 2		Al 1 Cl 3

The aluminium is balanced. To balance the chlorine, we put **2** in front of $AlCl_3$ and **3** in front of Cl_2.

Unbalanced:	Al	+	$3Cl_2$	→	$2AlCl_3$
	Al 1 Cl 6				Al 2 Cl 6

The aluminium is now unbalanced, therefore we must put a **2** in front of aluminium.

Balanced:	$2Al_{(s)}$	+	$3Cl_{2(g)}$	→	$2AlCl_{3(s)}$
	Al 2 Cl 6				Al 2 Cl 6

Cambridge IGCSE Chemistry Study and Revision Guide © David Besser

● Ionic equations

Starting from a balanced equation with state symbols, an ionic equation can be written using the following steps.

1 Anything with (aq) as a state symbol should be written as ions if it is

    a a dilute acid, e.g. $HCl_{(aq)}$ is written as $H^+_{(aq)}$ and $Cl^-_{(aq)}$

    b a metallic compound, e.g. $CuSO_{4(aq)}$ is written $Cu^{2+}_{(aq)}$ and $SO_4^{2-}_{(aq)}$

    c an ammonium salt, e.g. $(NH_4)_2SO_{4(aq)}$ is written $2NH_4^+_{(aq)}$ and $SO_4^{2-}_{(aq)}$.

2 Numbers in front of formulae in equations mean that everything after the number is multiplied, e.g. $2HNO_{3(aq)}$ is written as $2H^+_{(aq)}$ and $2NO_3^-_{(aq)}$.

3 The formulae of any substance with state symbols (s), (l), (g) are not written as ions, thus are not changed in an ionic equation.

4 Any ions which are the same on both sides, known as **spectator ions**, are cancelled.

● Sample exam-style question

Write the following balanced equation as an ionic equation.

$$Zn_{(s)} + CuSO_{4(aq)} \rightarrow ZnSO_{4(aq)} + Cu_{(s)}$$

Student's answer

$$Zn_{(s)} + Cu^{2+}_{(aq)} + \cancel{SO_4^{2-}_{(aq)}} \rightarrow Zn^{2+}_{(aq)} + \cancel{SO_4^{2-}_{(aq)}} + Cu_{(s)}$$

$SO_4^{2-}_{(aq)}$ are the same on both sides and are cancelled out.

$$Zn_{(s)} + Cu^{2+}_{(aq)} \rightarrow Zn^{2+}_{(aq)} + Cu_{(s)}$$

● Chemical calculations

Calculating relative molecular mass, M_r

Use the following relative atomic masses, A_r, to calculate the relative molecular masses of the compounds shown.

    H = 1   C = 12   N = 14   O = 16   Al = 27   S = 32   Pb = 207

- $CO_2 = 12 + (16 \times 2) = 44$
- $N_2O = (14 \times 2) + 16 = 44$
- $C_4H_{10} = (12 \times 4) + (1 \times 10) = 58$
- $Pb(NO_3)_2$. The recommended method is to multiply out the brackets, i.e. $PbN_2O_6 = 207 + (14 \times 2) + (16 \times 6) = 331$
- $Al_2(SO_4)_3 = Al_2S_3O_{12} = (27 \times 2) + (32 \times 3) + (16 \times 12) = 342$

Cambridge IGCSE Chemistry Study and Revision Guide © David Besser

How to calculate moles

Moles from masses

$$\text{Moles} = \frac{\text{mass (g)}}{\text{mass of one mole}}$$

Mass of 1 **mole** means relative atomic mass of any substance which only contains atoms. Relative molecular mass or relative formula mass should be used for all other substances.

Rearranging for mass:

$$\text{Mass (g)} = \text{moles} \times \text{mass of one mole}$$

Rearranging for mass of one mole:

$$\text{Mass of one mole} = \frac{\text{mass (g)}}{\text{moles}}$$

Please note that the mass must be in grams (g). If the mass is given in kilograms (kg), it must be multiplied by 1 000 or if it is given in tonnes, it must be multiplied by 1 000 000 to convert it into grams.

Moles from gas volumes

The volume of one mole of any gas is $24\,dm^3$ at room temperature and pressure.

$$24\,dm^3 = 24\,000\,cm^3$$

$$\text{Moles} = \frac{\text{volume}}{\text{volume of one mole}}$$

Rearranging for volume:

$$\text{Volume} = \text{moles} \times \text{volume of one mole}$$

The volume of the gas and the volume of one mole of gas must be in the same units when using these equations.

Moles from volumes and concentration of solutions

$$\text{Moles} = \text{volume}\,(dm^3) \times \text{concentration}\,(mol\,dm^{-3})$$

Rearranging for volume:

$$\text{Volume}\,(dm^3) = \frac{\text{moles}}{\text{concentration}\,(mol\,dm^{-3})}$$

Rearranging for concentration:

$$\text{Concentration}\,(mol\,dm^{-3}) = \frac{\text{moles}}{\text{volume}\,(dm^3)}$$

$\left.\phantom{\begin{array}{c}1\\1\\1\end{array}}\right\}$ volumes in dm^3

Correct units are very important in these equations. Because solutions are measured out using burettes and pipettes which are graduated in cm^3, the equations below may be more useful.

$$\text{Moles} = \frac{\text{volume}\,(cm^3) \times \text{concentration}\,(mol\,dm^{-3})}{1000}$$

Rearranging for volume:

$$\text{Volume}\,(cm^3) = \frac{\text{moles} \times 1000}{\text{concentration}\,(mol\,dm^{-3})}$$

Rearranging for concentration:

$$\text{Concentration}\,(mol\,dm^{-3}) = \frac{\text{moles} \times 1000}{\text{volume}\,(cm^3)}$$

$\left.\phantom{\begin{array}{c}1\\1\\1\end{array}}\right\}$ volumes in cm^3

Examiner's tip

It is a good idea to remember that both of these expressions have moles × 1000 on the top line.

Cambridge IGCSE Chemistry Study and Revision Guide © David Besser

Mole calculations using equations

The following calculations should be approached in the following order.

a Calculate any relative molecular masses, M_r, that are required.

b Calculate the number of moles of the substance where sufficient information is given to do so.

c Use the mole ratio in the equation to calculate the number of moles of the other substance.

d Use your answer to **c** to calculate either the

- mass or
- volume of gas or
- volume of solution or
- concentration of solution.

Examiner's tip

It is extremely important to show all the working out in calculations. If some correct working out is shown and the final answer is incorrect, you will still be awarded a considerable amount of credit.

● Worked examples

1 Calcium carbonate decomposes when it is heated according to the equation

$$CaCO_{3(s)} \rightarrow CaO_{(s)} + CO_{2(g)}$$

Calculate the mass of calcium oxide, CaO, that is produced when 20.0 g of calcium carbonate, $CaCO_3$, is heated until there is no further change. [3 marks]

Relative atomic masses, A_r: C = 12, O = 16, Ca = 40

a M_r: $CaCO_3 = 40 + 12 + (16 \times 3) = 100$

M_r: $CaO = 40 + 16 = 56$

b Moles of $CaCO_3 = 20 \div 100 = 0.20$ moles [1 mark]

c Mole ratio from the equation

1 mole $CaCO_3$: 1 mole CaO

0.20 moles $CaCO_3$: 0.20 moles of CaO [1 mark]

d Mass of CaO = moles × mass of 1 mole

$= 0.20 \times 56 = 11.2$ g [1 mark]

Examiner's tips

1 Because the question does not ask about carbon dioxide, CO_2, there is no need to calculate the relative molecular mass, M_r of carbon dioxide.

2 The final answer should always be expressed using correct units.

2 Calculate the volume of carbon dioxide at room temperature and pressure that is produced by heating 2.1 g of sodium hydrogen carbonate, $NaHCO_3$, according to the equation

$$2NaHCO_{3(s)} \rightarrow Na_2CO_{3(s)} + CO_{2(g)} + H_2O_{(l)}$$

A_r: Na = 23, H = 1, C = 12, O = 16

The volume of one mole of any gas is 24 dm^3 at room temperature and pressure. [4 marks]

a M_r: $NaHCO_3 = 23 + 1 + 12 + (16 \times 3) = 84$ [1 mark]

b Moles of $NaHCO_3 = 2.1 \div 84 = 0.025$ moles [1 mark] c Mole ratio from the equation

2 mole $NaHCO_3$: 1 mole CO_2

0.025 moles $NaHCO_3$: $0.025 \div 2 = 0.0125$ moles of CO_2 [1 mark]

d Volume of CO_2 = moles × volume of one mole of gas

$$= 0.0125 \times 24 = 0.3\,dm^3$$ [1 mark]

Examiner's tips

The question asks for the volume of carbon dioxide. It is a very common error to calculate the mass instead. Those who do this can achieve the first three marks as long as the working out is clearly shown.

It is very common for students not to use the mole ratio in the equation or to use it the wrong way round, i.e. 1:2 instead of 2:1. Again it is possible to score three marks out of four under these circumstances depending on how much correct working out is shown.

3 Calculate the volume of aqueous sodium hydroxide, $NaOH_{(aq)}$, of concentration $0.20\,mol\,dm^{-3}$ which would be required to neutralise exactly $25.0\,cm^3$ of dilute sulfuric acid, $H_2SO_{4(aq)}$, of concentration $0.25\,mol\,dm^{-3}$ according to the equation

$$2NaOH_{(aq)} + H_2SO_{4(aq)} \rightarrow Na_2SO_{4(aq)} + 2H_2O_{(l)}$$

a There are no masses involved in the question, so no M_r values have to be calculated.

b Moles of $H_2SO_4 = \dfrac{25.0 \times 0.25}{1000} = 6.25 \times 10^{-3}$ moles

c Mole ratio in equation

1 mole H_2SO_4 : 2 moles $NaOH$

$6.25 \times 10^{-3} \times 2 = 0.0125$ moles $NaOH$

d Volume of $NaOH = \dfrac{moles \times 1000}{concentration\ (mol\,dm^{-3})}$

$$= \dfrac{0.0125 \times 1000}{0.20} = 62.5\,cm^3$$

Examiner's tips

This question asks you to calculate the volume of a solution. Many candidates use the value of $24\,dm^3$ because they confuse the volume of a solution with the volume of a gas.

Many students calculate relative molecular masses, although there is no mention of mass in the question.

When calculating the number of moles of a solution, many use the equation

Moles = concentration × volume

which they often learn as

n = c × v

This equation can only be used if the volume is in dm^3, but in this case the volume is in cm^3 which means the factor of 1000 must be used.

31

4 240 dm³ of nitrogen, $N_{2(g)}$, reacts with excess hydrogen, $H_{2(g)}$, according to the equation

$$N_{2(g)} + 3H_{2(g)} \rightarrow 2NH_{3(g)}$$

a What would be the volume of ammonia, $NH_{3(g)}$, produced?

b What volume of hydrogen, $H_{2(g)}$, would react with the nitrogen?

All volumes are measured at room temperature and pressure.

The volume of one mole of any gas is 24 dm³ at room temperature and pressure.

a Ammonia

- There are no masses involved in the question, so no M_r values have to be calculated.

- Moles of nitrogen = 240 ÷ 24 = 10.0

- Mole ratio

  1 mole of nitrogen : 2 moles of ammonia

  10 moles of nitrogen: 2 × 10 = 20 moles of ammonia

- Volume of 20 moles of ammonia = 20 × 24 = 480 dm³

b Hydrogen

- Mole ratio

  1 mole of nitrogen : 3 moles of hydrogen

  10 moles of nitrogen: 3 × 10 = 30 moles of hydrogen

- Volume of 30 moles of hydrogen = 30 × 24 = 720 dm³

● Empirical formulae

The **empirical formula** is the smallest whole number ratio of the atoms of each element in a compound.

The empirical formula of a compound can be calculated if the masses of the elements that combine together are known. These masses can be expressed in units of mass (usually grams) or percentages by mass.

● Worked example

A compound contains the following percentage composition by mass: 26.7% carbon, 2.2% hydrogen and 71.1% oxygen.

A_r: C = 12, H = 1, O = 16

Percentage composition by mass means that 100 g of the compound contains 26.7 g of carbon, 2.2 g of hydrogen and 71.1 g of oxygen.

Cambridge IGCSE Chemistry Study and Revision Guide © David Besser

Method

Calculate the number of moles of atoms of each element.

- Carbon, $C = 26.7 \div 12 = 2.225$
- Hydrogen, $H = 2.2 \div 1 = 2.2$
- Oxygen, $O = 71.1 \div 16 = 4.44375$

Divide all the above by the smallest

- $C, 2.225 \div 2.2 = 1$
- $H, 2.2 \div 2.2 = 1$
- $O, 4.44375 \div 2.2 = 2$

Write down the empirical formula $= CHO_2$

Common errors

Common errors in determining the molecular formulae are:

- Using M_r instead of A_r e.g. using $O_2 = 32$ instead of $O = 16$ when calculating moles of atoms.
- Using atomic number instead of A_r when calculating moles of atoms.
- Over approximation e.g. if a compound contains manganese, Mn, and oxygen, O, and the number of moles of atoms is

    $Mn = 0.1$ and $O = 0.15$

Dividing both by the smallest

    $Mn = 0.1 \div 0.1 = 1$ and $O = 0.15 \div 0.1 = 1.5$

- Some candidates decide that 1.5 is approximately 1 and write the empirical formula as MnO. This is incorrect.
- Some candidates decide that 1.5 is approximately 2 and write the empirical formula as MnO_2. This is incorrect.
- The correct method is to multiply both $\times 2$, i.e. $1 \times 2 = 2$ and $1.5 \times 2 = 3$ and the empirical formula is Mn_2O_3.
- A number would have to be very close to a whole number (say 0.1 away) if such an approximation is to be made.

Examiner's tip

If dividing by the smallest does not produce a whole number in each case, multiply all the numbers by 2. If this still does not produce a whole number in each case, multiply all the numbers by 3. Continue until a whole number ratio is obtained.

● Molecular formulae

The **molecular formula** is the number of atoms of each element in one molecule of a substance.

Examples of molecular and empirical formulae are shown in Table 4.1.

Table 4.1 Examples of molecular and empirical formulae

Name	Molecular formula	Empirical formula
Butane	C_4H_{10}	C_2H_5
Hydrogen peroxide	H_2O_2	HO
Glucose	$C_6H_{12}O_6$	CH_2O
Benzene	C_6H_6	CH
Methane	CH_4	CH_4

Determination of molecular formulae from empirical formulae

It is possible to determine the molecular formula of a substance from its empirical formula alone, but only if the M_r of the substance is also known.

If the empirical formula of a compound is CH_2, the molecular formula of the compound can be expressed as $(CH_2)_n$, where n is a whole number.

If the M_r of the compound is 70, the M_r of $CH_2 = 12 + (1 \times 2) = 14$

$n = M_r$ of the compound $\div M_r$ of empirical formula

Therefore, $n = 70 \div 14 = 5$ and the molecular formula is $CH_2 \times 5 = C_5H_{10}$.

● Limiting reactants

When two substances are mixed, students usually assume that both substances will react completely and that neither is left over. This is possible, but it is also possible that too much of either substance is used, in which case one of the two substances will be left over at the end of the reaction. The substance that is all used up is called the **limiting reactant** and the other substance is said to be **in excess**.

● Worked example

5.6 g of iron, Fe, and 4.0 g of sulfur, S, are mixed together and heated. The equation is

$$Fe_{(s)} + S_{(s)} \rightarrow FeS(s)$$

Deduce which substance is the limiting reactant.

Moles of Fe = $5.6 \div 56 = 0.10$
Moles of S = $4.0 \div 32 = 0.125$
Mole ratio: 1 mole of Fe : 1 mole of S

Therefore, 0.10 mole of Fe reacts with 0.10 mole of S.
However, there are 0.125 moles of S. 0.125 is greater than 0.10, therefore some S is left over. S is in excess and Fe is the limiting reactant.

● Percentage yield

If the reactants shown in an equation are converted completely into the products, we say that the **percentage yield** is 100%. However, in some circumstances, yields are less than 100%.

● Worked example

0.60 g of magnesium ribbon, Mg, were burned in excess oxygen, O_2, according to the equation

$$2Mg_{(s)} + O_{2(g)} \rightarrow 2MgO_{(s)}$$

The mass of magnesium oxide, MgO, that was produced was found to be 0.80 g.
Calculate the percentage yield.

A_r: Mg = 24, O = 16
M_r of MgO = 24 + 16 = 40
Moles of Mg = $0.60 \div 24 = 0.025$
Mole ratio: 1 mole of Mg : 1 mole of MgO

Therefore, 0.025 mole of Mg : 0.025 mole of MgO
Mass of MgO = $0.025 \times 40 = 1.00$ g
Thus, if the yield was 1.00 g, the percentage yield would be 100%.
However, the yield is only 0.80 g.

Percentage yield = actual yield $\div$ 100% yield $\times$ 100%
Percentage yield = $0.80 \div 1.00 \times 100 = 80.0\%$

Cambridge IGCSE Chemistry Study and Revision Guide © David Besser

● Percentage purity

Naturally occurring substances are impure and contain less than 100% of a compound. An example is limestone which contains less than 100% of calcium carbonate, $CaCO_{3(s)}$. The percentage by mass of calcium carbonate in limestone is known as the **percentage purity**.

● Worked example

$1.00\,g$ of limestone is added to $100\,cm^3$ of $0.200\,mol\,dm^{-3}$ hydrochloric acid (an excess) (see equation 1).

Equation 1

$$CaCO_{3(s)} + 2HCl_{(aq)} \rightarrow CaCl_{2(aq)} + CO_{2(g)} + H_2O_{(l)}$$

The leftover acid was titrated and found to be neutralised by $24.8\,cm^3$, $0.100\,mol\,dm^{-3}$ of sodium hydroxide solution, NaOH (see equation 2).

Equation 2

$$NaOH_{(aq)} + HCl_{(aq)} \rightarrow NaCl_{(aq)} + H_2O_{(l)}$$

Moles of NaOH that reacted with excess acid $= \dfrac{24.8 \times 0.100}{1000}$

$$= 2.48 \times 10^{-3} \text{ moles}$$

Mole ratio in equation 2
1 mole NaOH reacts with 1 mole HCl

2.48×10^{-3} moles of NaOH react with 2.48×10^{-3} moles of HCl

Moles of HCl that was added to the limestone $= \dfrac{100 \times 0.2}{1000}$

$$= 0.02 \text{ moles}$$

Moles of HCl that reacted with calcium carbonate, $CaCO_3$
Moles of HCl added – moles HCl left over

$$0.02 - (2.48 \times 10^{-3}) = 0.01752 \text{ moles HCl}$$

Mole ratio in equation 2
2 moles HCl react with 1 mole $CaCO_3$
0.01752 moles HCl react with $\dfrac{0.01752}{2} = 8.76 \times 10^{-3}$ moles $CaCO_3$

M_r of $CaCO_3 = 40 + 12 + (16 \times 3) = 100$
Mass of $CaCO_3$ = moles × mass of 1 mole
$$= 8.76 \times 10^{-3} \times 100$$
$$= 0.876\,g$$

Percentage of $CaCO_3$ in limestone $= \dfrac{\text{mass of } CaCO_3}{\text{mass of limestone}} \times 100\%$

$$= \dfrac{0.876}{1.00} \times 100$$

$$= 87.6\%$$

35

Cambridge IGCSE Chemistry Study and Revision Guide © David Besser

Exam-style questions

The volume of one mole of any gas is 24 dm³ at room temperature and pressure.

A_r: H = 1; C = 12; O = 16; Al = 27; Cl = 35.5; K = 39; Ca = 40; Ti = 48

1 What mass of hydrogen gas is produced when 8.1 g of aluminium powder reacts with excess dilute hydrochloric acid according to the equation

$$2Al_{(s)} + 6HCl_{(aq)} \rightarrow 2AlCl_{3(aq)} + 3H_{2(g)}$$

[Total: 3 marks]

2 What volume of oxygen gas, $O_{2(g)}$, is produced at room temperature and pressure when 0.142 g of potassium superoxide, $KO_{2(s)}$, reacts with excess carbon dioxide, $CO_{2(g)}$, according to the equation

$$4KO_{2(s)} + 2CO_{2(g)} \rightarrow 2K_2CO_{3(s)} + 3O_{2(g)}$$

[Total: 3 marks]

3 What mass of calcium carbide, $CaC_{2(s)}$, is required to produce 120 cm³ of ethyne gas, $C_2H_{2(g)}$, by reaction with excess water according to the equation

$$CaC_{2(s)} + 2H_2O_{(l)} \rightarrow Ca(OH)_{2(aq)} + C_2H_{2(g)}$$

[Total: 3 marks]

4 20.0 cm³ of aqueous KOH neutralised 35.0 cm³ of dilute sulfuric acid, $H_2SO_{4(aq)}$, whose concentration was 0.20 mol dm⁻³. The equation is

$$2KOH_{(aq)} + H_2SO_{4(aq)} \rightarrow K_2SO_{4(aq)} + 2H_2O_{(l)}$$

Calculate the concentration of the aqueous potassium hydroxide, $KOH_{(aq)}$, in

a mol dm⁻³

b g dm⁻³   [Total: 4 marks]

5 A compound has composition by mass which is 54.5% carbon, 9.1% hydrogen and 36.4% oxygen.

The M_r of the compound = 44. Calculate the

a empirical formula

b molecular formula of the compound.   [Total: 4 marks]

6 When 0.38 g of titanium (IV) chloride, $TiCl_{4(s)}$, reacted with excess sodium, the reaction produced 0.024 g of titanium, $Ti_{(s)}$. The equation is

$$TiCl_{4(s)} + 4Na_{(s)} \rightarrow Ti_{(s)} + 4NaCl_{(s)}$$

Calculate the percentage yield of titanium.   [Total: 4 marks]

5 — Electricity and chemistry

Key objectives

By the end of this section, you should

- be able to define electrolysis, electrolyte and electrode
- be able to describe the products of electrolysis and state the observations made when various electrolytes conduct electricity
- be able to describe the electroplating of metals and outline the uses of electroplating
- be able to describe and explain the reasons why copper and (steel-cored) aluminium are used in cables and why plastics are used as insulators

- be able to construct ionic half-equations for reactions at the cathode
- understand the reasons why different substances are conductors, insulators and electrolytes
- be able to describe production of energy from simple cells
- be able to describe in outline the manufacture of aluminium from pure aluminium oxide in molten cryolite (see Chapter 10)
- be able to describe in outline the production of chlorine, hydrogen and sodium hydroxide from concentrated aqueous sodium chloride
- know what the products of electrolysis of aqueous copper(II) sulfate are with both carbon electrodes and copper electrodes, and relate this to the refining of copper (see Chapter 10).

● Key terms

Electrolysis	The process by which an ionic compound, when molten or in aqueous solution, is chemically changed by the passage of an electric current
Electrolyte	A liquid which is chemically changed by an electric current
Electrodes	The conducting rods by which the electric current enters and leaves the electrolyte
Anode	The positive (+) electrode
Cathode	The negative (−) electrode

Substances that conduct electricity can be subdivided into conductors and electrolytes (Table 5.1).

Table 5.1 Differences between conductors and electrolytes

	Conductors	Electrolytes
Physical state	Solid	Liquid
Differences	Conduct electricity, but are not chemically changed by the electric current. The only change undergone by conductors is that they become hot, which is a physical change	Conduct electricity and are chemically changed by the electric current. The products of chemical change are formed at the electrodes
Examples	All metallic elements and alloys Graphite and graphene	Molten ionic compounds Aqueous solutions containing ions
Particles responsible for conduction	Moving (mobile) electrons	Moving ions

Examiner's tip

The terms electrolysis, electrolyte and electrode all begin with the same letter. Make sure you know the difference between these terms.

● Electrolytes

Electrolytes must be in the liquid state. Solid ionic compounds, e.g. sodium chloride, do not conduct electricity in the solid state because although they contain ions, the ions are held together by strong forces of attraction in the giant ionic lattice. Because the ions are not moving, solid sodium chloride does not conduct electricity.

Cambridge IGCSE Chemistry Study and Revision Guide © David Besser

There are two ways to make ionic solids into electrolytes:

1 Heat until the ionic solid melts. This requires a large amount of heat energy, because ionic compounds have high melting points (see Chapter 3). Molten ionic compounds are electrolytes because the ions are moving in the liquid state.

2 Dissolve the ionic solid in water. An aqueous solution of an ionic compound also contains moving ions.

When molten ionic compounds and aqueous solutions of ionic compounds conduct electricity, the positive ions (cations) move to the **cathode** (–) and the negative ions (anions) move to the **anode** (+).

At the cathode, positive ions gain electrons and are reduced (see Chapter 7), for example

$$2H^+ + 2e^- \rightarrow H_2$$

At the anode, negative ions lose electrons and are oxidised (see Chapter 7), for example

$$2Cl^- \rightarrow Cl_2 + 2e^-$$

Thus, the products of **electrolysis** are formed at the **electrodes**.

When ions lose their charge to form atoms or molecules they are said to be discharged.

● Practical electrolysis

Electrolysis can be carried out in school laboratories using the apparatus shown in Figure 5.1.

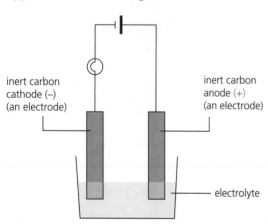

Figure 5.1 The important terms used in electrolysis

The electrolyte is placed in a crucible, if it is a solid that has to be heated until its melting point, or a beaker, if it is a liquid at room temperature.

Figure 5.2 shows how gaseous products can be collected during electrolysis, as in the electrolysis of dilute sulfuric acid.

The products that form during electrolysis are summarised in Table 5.2.

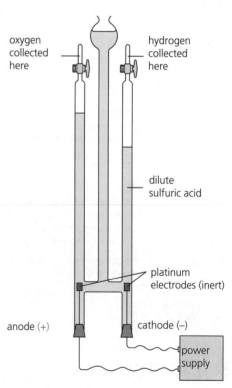

Figure 5.2 A Hofmann voltameter used to electrolyse water

Table 5.2 Summary of products formed during electrolysis

Type of electrolyte	Products at anode (+)	Products at cathode (−)
Molten ionic compound	Non-metallic element	Metallic element
Aqueous solutions containing ions	Either oxygen gas or halogen (chlorine, bromine or iodine) from any concentrated aqueous solution of a halide (chloride, bromide or iodide)	Either hydrogen gas or metallic element below hydrogen in the reactivity series, e.g. copper

Examiner's tip

Remember that molten ionic compounds produce a non-metallic element at the anode and a metallic element at the cathode.

Aqueous solutions produce oxygen or a halogen at the anode and hydrogen or a metal at the cathode .The hydrogen and oxygen come from the water that is contained in the aqueous solution.

Some examples of products of the electrolysis of different electrolytes, using carbon or platinum (inert electrodes), are shown in Table 5.3.

Table 5.3 Examples of products of the electrolysis of different electrolytes, using inert electrodes

	Product at anode (+)	Observations at anode (+)	Ionic half-equation for reaction at anode (+)	Product at cathode (−)	Observations at cathode (−)	Ionic half-equation for reaction at cathode (−)
Molten sodium chloride, $NaCl_{(l)}$	Chlorine	Bubbles of green gas	$2Cl^- \rightarrow Cl_2 + 2e^-$	Sodium	Grey metal coating	$Na^+ + e^- \rightarrow Na$
Concentrated aqueous sodium chloride, $NaCl_{(aq)}$	Chlorine	Bubbles of green gas	$2Cl^- \rightarrow Cl_2 + 2e^-$	Hydrogen	Bubbles of colourless gas	$2H^+ + 2e \rightarrow H_2$
Molten lead bromide, $PbBr_{2(l)}$	Bromine	Bubbles of brown gas	$2Br^- \rightarrow Br_2 + 2e^-$	Lead	Grey metal coating	$Pb^{2+} + 2e^- \rightarrow Pb$
Concentrated hydrochloric acid, $HCl_{(aq)}$	Chlorine	Bubbles of green gas	$2Cl^- \rightarrow Cl_2 + 2e^-$	Hydrogen	Bubbles of colourless gas	$2H^+ + 2e \rightarrow H_2$
Dilute sulfuric acid, $H_2SO_{4(aq)}$	Oxygen	Bubbles of colourless gas	$4OH^- \rightarrow 2H_2O + O_2 + 4e^-$	Hydrogen	Bubbles of colourless gas	$2H^+ + 2e \rightarrow H_2$
Aqueous copper(ɪɪ) sulfate, $CuSO_{4(aq)}$	Oxygen	Bubbles of colourless gas	$4OH^- \rightarrow 2H_2O + O_2 + 4e^-$	Copper	Pink metal coating	$Cu^{2+} + 2e^- \rightarrow Cu$

Examiner's tip

Make sure you remember that

* Very reactive metals that react with cold water (such as potassium, sodium and calcium) cannot be produced by electrolysis of aqueous solutions.

* During the electrolysis of all aqueous solutions containing positive ions of a metal above hydrogen in the reactivity series, hydrogen is produced at the cathode as opposed to the metallic element. The extraction of these metals by electrolysis can only be carried out using a molten electrolyte (see extraction of aluminium, p. 73, in Chapter 10).

* Aqueous solutions of acids always produce hydrogen at the cathode (−).This is because the positive H^+ ion is common to both the acidic substance and water.

Cambridge IGCSE Chemistry Study and Revision Guide © David Besser

Electrolysis of copper(II) sulfate

If aqueous copper(II) sulfate is electrolysed using carbon or platinum electrodes (inert electrodes), the products are copper at the cathode and oxygen at the anode (see Table 5.3).

However, if the anode is made of copper, oxygen is not produced at the anode. Instead, the copper anode goes into solution as positive ions:

$$Cu_{(s)} \rightarrow Cu^{2+}_{(aq)} + 2e^-$$

Eventually, the $Cu^{2+}_{(aq)}$ ions reach the cathode where the reverse reaction occurs and copper metal is formed:

$$Cu^{2+}_{(aq)} + 2e^- \rightarrow Cu_{(s)}$$

This electrolytic process is used in the refining (purification) of copper which is carried out on a large scale (see Chapter 10).

- The anode is impure copper
- The electrolyte is aqueous copper(II) sulfate
- The cathode is pure copper.

The copper at the anode goes into solution as positive Cu^{2+} ions. Impurities either go into solution as positive ions or fall off the anode and are deposited at the bottom of the container. Pure copper forms at the cathode. None of the other metallic ions are discharged at the cathode.

Pure copper is essential when copper is being used as an electrical conductor, as in electrical wiring. Impurities in the copper decrease its electrical conductivity considerably.

Electrolysis of concentrated aqueous sodium chloride

Sodium chloride (common salt) is first mined and then electrolysed as a concentrated aqueous solution. Electrolysis of concentrated aqueous sodium chloride can be carried out in the laboratory or on an industrial scale.

The ions that are present in the solution are

$Na^+_{(aq)}$ and $Cl^-_{(aq)}$ (from $NaCl_{(aq)}$)

(Small amounts of) $H^+_{(aq)}$ and $OH^-_{(aq)}$ (from $H_2O_{(l)}$)

At the anode, hydrogen gas is produced $2H^+ + 2e^- \rightarrow H_2$.
The $Na^+_{(aq)}$ remain in the solution.
At the cathode, chlorine gas is produced $2Cl^- \rightarrow Cl_2 + 2e$.
The $OH^-_{(aq)}$ remain in the solution.
Because the electrolyte contains $Na^+_{(aq)}$ and $OH^-_{(aq)}$, the electrolyte has changed into aqueous sodium hydroxide, $NaOH_{(aq)}$. The three products of electrolysis have several industrial uses.

Product	Hydrogen	Chlorine	Sodium hydroxide
Use	To make ammonia Fuel in fuel cells	Water treatment (to kill bacteria) Manufacture of PVC	Extraction of aluminium Manufacture of soap

● Electroplating

Electroplating is another electrolytic process that can be carried out in a school laboratory or on a large scale. Electroplating means coating an object with a thin layer of a metal.

The purpose is

- to improve appearance
- to prevent corrosion, e.g. rusting.

Electroplating is carried out using

- the plating metal as the anode
- the object to be plated as the cathode
- an aqueous solution containing ions of the plating metal as the electrolyte.

In the example of silver plating shown in Figure 5.3, the silver anode goes into solution as silver ions.

$$Ag_{(s)} \rightarrow Ag^+_{(aq)} + e^-$$

The silver ions in the electrolyte are discharged at the cathode.

$$Ag^+_{(aq)} + e^- \rightarrow Ag_{(s)}$$

The silver ions that are released at the anode replace those that are discharged. The silver produced at the cathode electroplates the spoon.

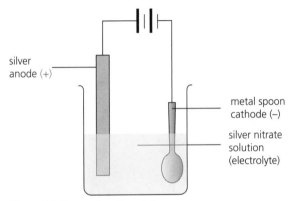

silver anode (+)

metal spoon cathode (−)

silver nitrate solution (electrolyte)

Figure 5.3 Process for silver plating

● Extraction of metals

Reactive metals are extracted by electrolysis of molten ionic compounds (see extraction of aluminium, pp. 73–4, Chapter 10).

Unreactive metals are extracted by electrolysis of aqueous solutions (see refining of copper, pp. 75–6, Chapter 10).

● Uses of metals, plastics and ceramics

Aluminium is used in steel-cored cables, because it

- is unreactive due to a coating of aluminium oxide
- is a good conductor of electricity
- has a low density.

Copper is used in electric cables and wires due to very high electrical conductivity.

Plastics and ceramics do not conduct electricity and therefore are used as insulators in the power supply industry.

● Cells

Electrolysis uses electrical energy to carry out chemical reactions.

In **cells**, chemical reactions are used to produce electrical energy. If two dissimilar metals are placed in an electrolyte, electrical energy is produced. This is the principle of the battery. Such cells can be set up in order to put metals in order of reactivity (see Chapter 10).

Fuel cells have a fuel, such as hydrogen or ethanol. The fuel reacts with oxygen in order to generate electricity (see Chapter 6).

Cambridge IGCSE Chemistry Study and Revision Guide © David Besser

Exam-style questions

1   Complete the following table to show the products of electrolysis using carbon electrodes.

Electrolyte	Name of product at anode (+)		Name of product at cathode (−)	
Molten potassium bromide		[1 mark]		[1 mark]
Aqueous potassium bromide		[1 mark]		[1 mark]
Molten lead iodide		[1 mark]		[1 mark]
Aqueous copper(II) chloride		[1 mark]		[1 mark]
Aqueous sodium sulfate		[1 mark]		[1 mark]

[Total: 10 marks]

2   A student wanted to electroplate a knife with nickel. What should the student use as

a   the anode

b   the electrolyte

c   the cathode?

[Total: 3 marks]

3   A student carries out electrolysis of concentrated aqueous potassium iodide in a beaker using carbon electrodes.

a   Name the product at the anode.   [1 mark]

b   Write an ionic half-equation for the reaction occurring at the cathode.   [1 mark]

c   State the type of reaction occurring at the anode.   [1 mark]

d   State the name of the solution left in the beaker when the electrolysis has finished.   [1 mark]

e   Name the type of particles that are responsible for the conduction of electricity in the conducting wire.   [1 mark]

f   Name the type of particles that are responsible for the conduction of electricity in the electrolyte.   [1 mark]

[Total: 6 marks]

6 Chemical energetics

● Key terms

Exothermic reaction	A reaction in which there is an overall transfer of energy to the surroundings
Endothermic reaction	A reaction in which there is an overall gain of energy from the surroundings
Bond energy	The amount of energy required to break one mole of covalent bonds in gaseous molecules

Examples of **exothermic reactions** are

- combustion of fuels such as alkanes and hydrogen gas; fuels release heat energy when they burn in air or oxygen.
- respiration (see Chapter 13).

Examples of **endothermic reactions** are

- photosynthesis (see Chapter 13)
- thermal decomposition (see Chapters 10 and 13).

● Energy level diagrams

Exothermic and endothermic reactions can be represented by **energy level diagrams**. These diagrams show the energy of the reactants and products as a reaction progresses.

● Exothermic reactions

The complete combustion of fuels, such as methane, is an example of an exothermic reaction (Figure 6.1). In any exothermic reaction, the products have less energy than the reactants because energy has been given out to the surroundings.

Figure 6.1 Energy level diagram for the complete combustion of methane

Cambridge IGCSE Chemistry Study and Revision Guide © David Besser

● Endothermic reactions

The thermal decomposition of calcium carbonate is an example of an endothermic reaction (Figure 6.2). In any endothermic reaction, the products have more energy than the reactants because energy has been taken in from the surroundings.

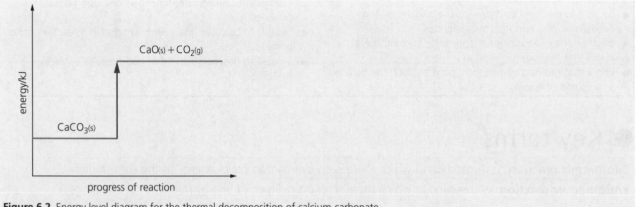

Figure 6.2 Energy level diagram for the thermal decomposition of calcium carbonate

● Fuels

Fossil fuels (petroleum, natural gas and coal) all release energy when they burn.

Uranium 235 (^{235}U) is a source of nuclear energy. It has the advantage of not causing global warming as no greenhouse gases are released when it is used as a fuel, but there are safety issues concerning the adverse effects of radiation.

Hydrogen also releases energy when it burns. An advantage of using hydrogen as a fuel as an alternative to fossil fuels is that water is the only product; carbon dioxide (a greenhouse gas) is not produced.

Hydrogen can also be used, along with oxygen, in a **fuel cell** to release electrical energy.

● Fuel cells

In hydrogen fuel cells, the overall reaction is the same as when hydrogen is burned in air or oxygen.

$$2H_{2(g)} + O_{2(g)} \rightarrow 2H_2O_{(l)}$$

Fuel cells operate in acidic or alkaline conditions. An alkaline hydrogen fuel cell is shown in Figure 6.3.

One advantage of carrying out this reaction in a fuel cell is that fuel cells are much more efficient than internal combustion engines, which means there is much less energy loss.

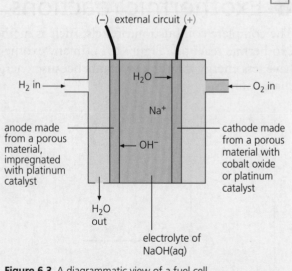

Figure 6.3 A diagrammatic view of a fuel cell

Cambridge IGCSE Chemistry Study and Revision Guide © David Besser

Bond energies

Most chemical reactions involve breaking of covalent bonds in the reactants and formation of new covalent bonds in the products.

- Breaking of bonds is an endothermic process (energy is taken in).

- Formation of bonds is an exothermic process (energy is given out).

Because the amount of energy put in to break bonds is very unlikely to be equal to the amount of energy given out when new bonds are formed, most reactions are either endothermic or exothermic.

An example is the reaction between gaseous hydrogen and gaseous chlorine to form gaseous hydrogen chloride. The equation is

$$H_{2(g)} + Cl_{2(g)} \rightarrow 2HCl_{(g)}$$

Bond energy is the amount of energy required to break 1 mole of covalent bonds in gaseous molecules. It is numerically equal to the amount of energy given out when new bonds form in gaseous molecules.

When covalent bonds in molecules are broken, the molecules change into atoms. The atoms then join together to form new molecules.

Bond energies are shown in Table 6.1.

Table 6.1 Bond energies (kJ/mole)

Bond	Bond energy (kJ/mole)
H–H	435
Cl–Cl	242
H–Cl	432

The equation can be written to show the structure of the molecules:

$$H-H + Cl-Cl \rightarrow 2H-Cl$$

Energy taken in to break bonds	Energy given out when bonds form
H–H = 435 kJ	**2** × H–Cl = **2** × 432 = 864
Cl–Cl = 242 kJ	
Total energy put in: 435 + 242 = 677 kJ	Total energy given out = 864 kJ

Because 864 is a larger number than 677, this means that more energy is given out when the bonds form in the products than the energy that has to be put in to break the bonds in the reactants. Therefore, the reaction is exothermic and the overall energy change is

$$864 - 677 = 187 \text{ kJ/mole}$$

which means that when 1 mole of gaseous H_2 molecules react with 1 mole of gaseous Cl_2 molecules to form 2 moles of gaseous HCl molecules, 187 kJ of energy are given out to the surroundings.

This can also be shown on an energy level diagram.

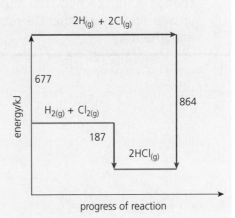

Figure 6.4 Energy level diagram

Cambridge IGCSE Chemistry Study and Revision Guide © David Besser

Exam-style questions

1 Propane burns in excess oxygen to form carbon dioxide and water according to equation below. (Note that the table of bond energies will be required to answer the question.)

$$C_3H_{8(g)} + 5O_{2(g)} \rightarrow 3CO_{2(g)} + 4H_2O_{(g)}$$

Bond	Bond energy kJ/mol
C–C	347
C–H	435
O=O	497
C=O	803
O–H	464

Calculate the overall energy change occurring when 1 mole of $C_3H_{8(g)}$ reacts with 5 moles of $O_{2(g)}$ to form 3 moles of $CO_{2(g)}$ and 4 moles of $H_2O_{(g)}$ by using the following steps:

a Draw the structures of all the molecules on both sides of the equation. Show all the atoms and all the bonds.   [2 marks]

b Write down the number of moles of each type of bond that have to be broken in the reactants. (Remember to consider the number of moles of both reactants in the equation.)   [2 marks]

c Calculate the total amount of energy that has to be put in to break all the bonds in the reactants in (b).   [1 mark]

d Write down the number of moles of each type of bond that have to be formed in the products. (Remember to consider the number of moles of both products in the equation.)   [2 marks]

e Calculate the total amount of energy that is given out when all the bonds in the products in (d) are formed.   [1 mark]

f Use your answers to (c) and (e) to calculate the overall energy change in the reaction. You must show the correct units in your answer and say whether the reaction is exothermic or endothermic.   [3 marks]

Cambridge IGCSE Chemistry Study and Revision Guide © David Besser

7 Chemical reactions

● Key terms

Catalyst	A substance which increases the rate of a chemical reaction. The catalyst is chemically unchanged at the end of the reaction
Enzyme	A biological catalyst. Enzymes are protein molecules
Oxidation	Gain of oxygen or loss of hydrogen
Reduction	Loss of oxygen or gain of hydrogen
Oxidising agent	A substance that oxidises another substance in a redox reaction. An electron acceptor
Reducing agent	A substance that reduces another substance in a redox reaction. An electron donor
Oxidation	Loss of electrons
Reduction	Gain of electrons

Physical changes are changes in which new chemical substances *are not* produced. Changes in state, that is melting, boiling, evaporation, condensation, sublimation and freezing (see Chapter 1), and separation of mixtures, e.g. filtration, distillation, fractional distillation, chromatography and crystallisation (see Chapter 2), are examples of physical changes.

Chemical changes or chemical reactions are changes in which new chemical substances *are* produced. Decomposition, electrolysis, respiration, photosynthesis, redox, neutralisation, cracking, addition, substitution, polymerisation and combustion are examples of chemical changes. Chemical equations can always be used to represent chemical changes.

The term **physical properties** of a substance refers to properties of a substance that can be measured and that involve physical changes. Examples are melting point, boiling point and density.

The term **chemical properties** of a substance refers to properties of a substance which involve chemical changes. Examples are the things that substances react with and details of such reactions.

It can be said that a physical property of metals is that they all conduct electricity, whereas a chemical property of metals is that (some of them) react with acids to produce a salt and hydrogen.

Examiner's tip

Students should ensure that they are aware of the differences between *physical properties* and *chemical properties* and that they know examples of both for different types of substance.

● Rate of reaction

The **rate** of a chemical reaction can be determined by measuring either how the amount of one of the reactants decreases with time or how the amount of one of the products increases with time.

Common errors

- Students commonly refer to **catalysts** being able to alter the speed of a chemical reaction. This does not specifically mean that a catalyst increases the rate, because altering the speed suggests that catalysts may decrease the rate, which is not the case.
- Another common error is to suggest that catalysts do not take part in a reaction. This is not the case, because increasing the rate suggests that catalysts have a considerable part to play.
- The term 'biocatalyst' is not recognised as having the same meaning as biological catalyst.

● Experimental investigations

Reactions in which solids react with liquids to produce gases among other products are commonly investigated in this chapter.

Experiment 1

An example is

$$Zn_{(s)} + H_2SO_{4(aq)} \rightarrow ZnSO_{4(aq)} + H_{2(g)}$$

The reaction between a known excess of zinc granules and $50.0\,cm^3$ of $0.10\,mol\,dm^{-3}$ dilute sulfuric acid was investigated by a student. The apparatus shown in Figure 7.1 was used.

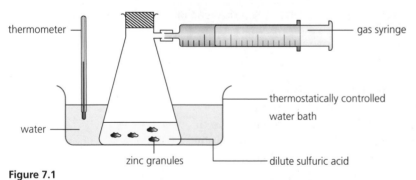

thermometer — gas syringe

thermostatically controlled water bath

water —

zinc granules — dilute sulfuric acid

Figure 7.1

Cambridge IGCSE Chemistry Study and Revision Guide © David Besser

The temperature is kept at 25 °C by using a thermostatically controlled water bath. The volume of hydrogen produced can be measured at regular time intervals. A graph is then plotted (Figure 7.2).

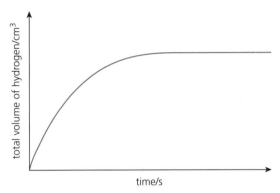

Figure 7.2 The volume of hydrogen produced against time for Experiment 1

The graph is steepest at the start which means that this is when the rate of reaction is fastest. It then becomes less steep which means that the reaction becomes slower. Eventually the graph levels out, which means that no more hydrogen gas is released and the rate of reaction is zero.

● Collision theory

In any reaction of the type

$$A + B \rightarrow C$$

particles of A and B must collide with each other if they are to produce product C. There are two types of collision: successful and unsuccessful.

In an unsuccessful collision, particles of A and B merely bounce off each other and remain as A and B.

However, in a successful collision, particles of A and B collide and change into C.

Collisions are only successful if the reacting particles contain at least a minimum amount of energy called the **activation energy**.

The rate of a chemical reaction depends on the number of successful collisions occurring in any given time.

If a change is made to increase the number of collisions, this automatically leads to an increase in the number of successful collisions because a certain proportion of all collisions are always successful.

In the reaction investigated

$$Zn_{(s)} + H_2SO_{4(aq)} \rightarrow ZnSO_{4(aq)} + H_{2(g)}$$

the ionic equation is

$$Zn_{(s)} + 2H^+_{(aq)} \rightarrow Zn^{2+}_{(aq)} + H_{2(g)}$$

This shows that collisions must take place between zinc atoms and hydrogen ions for the reaction to occur. In Experiment 1, the rate of reaction is fastest at the start because this is where the concentration of hydrogen ions is highest, which means that the number of collisions between hydrogen ions and zinc atoms in any given time is most frequent at the start.

The rate of reaction then decreases because as the concentration of hydrogen ions decreases, collisions occur less frequently.

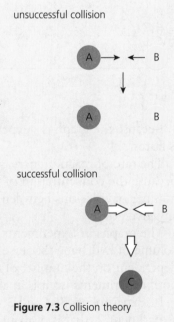

unsuccessful collision

successful collision

Figure 7.3 Collision theory

When all the sulfuric acid is used up, the concentration of hydrogen ions becomes zero, therefore there are no more collisions and the rate becomes zero.

The student then carried out experiments on the same reaction, changing only one variable in each case. The changed variable is highlighted.

Experiment	Temperature/°C	Catalyst	Sulfuric acid, $H_2SO_{4(aq)}$	Zinc, $Zn_{(s)}$
1	25	None	50.0 cm³ of 0.10 mol dm⁻³	Granules
2	25	None	**25.0 cm³ of 0.20 mol dm⁻³**	Granules
3	25	None	50.0 cm³ of 0.10 mol dm⁻³	**Powder**
4	**50**	None	50.0 cm³ of 0.10 mol dm⁻³	Granules
5	25	**A few drops of aqueous copper(ɪɪ) sulfate**	50.0 cm³ of 0.10 mol dm⁻³	Granules

Examiner's tip

In reactions between gases, it is possible to make similar statements about the concentration of a gas. However, it is more likely that gases are referred to in terms of pressure rather than concentration. The higher the pressure exerted by a gas, the closer together are the gaseous molecules and the more frequent will be their collisions with each other.

Experiment 2

The concentration of sulfuric acid is doubled in Experiment 2 but the volume of sulfuric acid is halved which means that the number of moles of sulfuric acid is the same.

Figure 7.4 shows the graph that was plotted in Experiment 2 with the graph for Experiment 1.

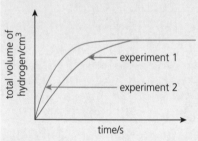

Figure 7.4

Because the graph is steeper at the start, this is where the rate of reaction is fastest.

The rate of reaction increases as the concentration increases. This is because the concentration of hydrogen ions is higher, which means that the number of collisions between hydrogen ions and zinc atoms is also higher in any given time.

The graphs in Experiments 1 and 2 (Figure 7.4) level off at the same volume of hydrogen, because the volume of hydrogen given off is only dependent on the number of moles of sulfuric acid, which is the same in both experiments (as it is in all five experiments).

In Experiments 3, 4 and 5, the graph would be the same shape as in Experiment 2, that is, it would be steeper at the start and level off at the same volume of hydrogen.

Experiment 3

Using zinc powder we are decreasing particle size/increasing surface area. Collisions can only occur on the surface of the zinc. With a smaller particle size, there are more zinc atoms available to collide with the hydrogen ions. More collisions occurring in any given time means that more successful collisions occur in any given time and therefore a faster rate of reaction occurs.

Decreased particle size and increased rate of reaction can lead to hazards, for example, in flour mills and coal mines. Flour dust and coal dust (both having extremely large surface areas) have been known to react explosively with oxygen in the air when a spark has been created by machinery.

Experiment 4

At higher temperature, the reacting particles have more energy. This means that the particles move faster and collide more often in any given time.

However, there will be more particles with energy greater than or equal to the activation energy. Therefore, more collisions will be successful collisions in any given time. This is the main reason why rates of reaction are faster at higher temperatures.

Experiment 5

Aqueous copper(II) sulfate acts as a catalyst in this reaction.

Catalysts lower the activation energy of a reaction. This means that there are more particles with energy greater than or equal to the activation energy. Therefore, more collisions are successful collisions in any given time and the rate of reaction is faster.

Activation energy is the amount of energy that has to be supplied to reactants to make a reaction occur. The lowering of activation energy in a catalysed reaction can be shown in the energy profile (Figure 7.5). As can be seen, using a catalyst has no effect on the overall energy change of a reaction but lowers the activation energy, thus increasing the rate of reaction.

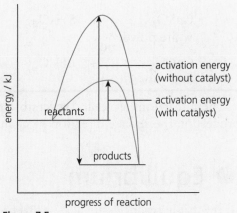

Figure 7.5

● Photochemical reactions

Photochemical reactions require light in order to occur. Furthermore, the rate of such reactions increases when light intensity increases.

Examples of photochemical reactions are:

- photosynthesis in which carbon dioxide and water react in the presence of sunlight and chlorophyll to produce glucose and oxygen (see Chapter 13)
- substitution reaction of alkanes with chlorine (see Chapter 14) which requires UV light
- photochemical decomposition of silver salts in photography. When aqueous silver nitrate is added to an aqueous solution containing halide ions (Cl^-, Br^- or I^-), a precipitate of a silver halide occurs, e.g.

$$Ag^+_{(aq)} + Br^-_{(aq)} \rightarrow AgBr_{(s)}$$

When the silver bromide is exposed to light it decomposes into its elements and the precipitate darkens as it gradually decomposes into silver and bromine. Silver bromide is used in photographic films because it darkens when exposed to light.

$$2AgBr_{(s)} \rightarrow 2Ag_{(s)} + Br_{2(g)}$$

The rate of this reaction increases as light intensity increases.

● Reversible reactions

Some reactions can be reversed by changing the conditions. If crystals of hydrated copper(II) sulfate and hydrated cobalt(II) chloride are heated, they change colour as they lose their water of crystallisation and become anhydrous salts.

$$CuSO_4.5H_2O_{(s)} \rightarrow CuSO_{4(s)} + 5H_2O_{(g)}$$
blue crystals   white powder

$$CoCl_2.6H_2O_{(s)} \rightarrow CoCl_{2(s)} + 6H_2O_{(g)}$$
pink crystals   blue powder

However, in both cases, the reactions can be made to proceed in the reverse directions by adding water to the anhydrous salts in which case the crystals form again, as can be seen by the reverse colour change.

$$CuSO_{4(s)} + 5H_2O_{(l)} \rightarrow CuSO_4.5H_2O_{(s)}$$
white powder   blue crystals

$$CoCl_{2(s)} + 6H_2O_{(l)} \rightarrow CoCl_2.6H_2O_{(s)}$$
blue powder   pink crystals

These reactions are called **reversible reactions**. They can be made to proceed in the reverse direction by changing the conditions.

● Equilibrium

If reversible reactions are allowed to proceed in a closed container, they reach a state that is known as **chemical equilibrium**.

If a mixture of hydrogen and iodine gases is heated in a closed container, the hydrogen reacts with the iodine to produce hydrogen iodide:

$$H_{2(g)} + I_{2(g)} \rightarrow 2HI_{(g)}$$

This is called the forward reaction.

As soon as hydrogen iodide molecules are formed, they start to decompose into hydrogen and iodine:

$$2HI_{(g)} \rightarrow H_{2(g)} + I_{2(g)}$$

This is called the reverse or backward reaction.

Therefore, two reactions are occurring in the same container at the same time. Furthermore, one reaction is the reverse of the other. This can be shown by the following expression:

$$H_{2(g)} + I_{2(g)} \rightleftharpoons 2HI_{(g)} \quad \text{Reactants} \rightleftharpoons \text{Products}$$

The forward reaction starts off quickly and the rate decreases as the concentrations of hydrogen and iodine decrease.

The backward reaction starts off slowly and the rate increases as the concentration of hydrogen iodide increases.

Eventually, both rates become equal. The system is then in a state of chemical equilibrium. When this occurs, reactants and products are all being used up and produced at the same rate and therefore their concentrations become constant.

Characteristics of equilibrium systems

Equilibrium can only occur in a closed system (closed container), in which no substances can escape to the outside or enter from the outside.

- The rate of the forward reaction is equal to the rate of the reverse reaction.

- The concentrations of all reactants and products become constant.

Examiner's tip

When asked to describe the characteristics of an equilibrium system, students make two very common errors. They often state:

1 *The forward reaction is equal to the reverse reaction.* Without using the word 'rate', this is a meaningless statement.

2 a *The amounts of reactants and products become constant.* In this case, the word 'amounts' must be replaced by 'concentrations'.
   b *The concentrations of products and reactants become equal.* This is incorrect. 'The concentrations of products and reactants become **constant**, which means that they stop changing', is the correct statement.

● Effects of changing the conditions of an equilibrium system

The following equation represents the equilibrium which occurs in the Haber process:

$$N_{2(g)} + 3H_{2(g)} \rightleftharpoons 2NH_{3(g)}$$

The forward reaction is exothermic.
This means that $N_{2(g)} + 3H_{2(g)} \rightarrow 2NH_{3(g)}$ is an exothermic reaction.
Therefore, $2NH_{3(g)} \rightarrow N_{2(g)} + 3H_{2(g)}$ is an endothermic reaction.
The equation shows that there are four gaseous molecules on the left-hand side of the $\rightleftharpoons$ sign and two gaseous molecules on the right-hand side of the $\rightleftharpoons$ sign.

Table 7.1 The conditions of this equilibrium system and their results

Change	Effect on equilibrium position	Result in this example
Increase the concentration of reactants	Shifts to the right	Concentration of products increases
Increase the concentration of products	Shifts to the left	Concentration of reactants increases
Increase the total pressure	Shifts in the direction of fewer molecules	Concentration of products increases
Increase temperature	Shifts in endothermic direction	Concentration of reactants increases
Add a catalyst	Increases the rate of both forward and reverse reactions, but does not shift the equilibrium	No change

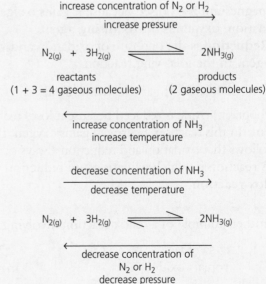

Figure 7.6

Decreases in concentrations, pressure and temperature have the opposite effect to increases. This is summarised in Figure 7.6.

Cambridge IGCSE Chemistry Study and Revision Guide © David Besser

Examiner's tips

When students are asked about changing temperature and pressure on an equilibrium system, they often demonstrate confusion between rate of reaction and equilibrium. This is shown in the sample question below.

● Sample exam-style question

What happens to the position of equilibrium in the Haber process reaction, i.e. $N_{2(g)} + 3H_{2(g)} \rightleftharpoons 2NH_{3(g)}$, when the temperature increases? The forward reaction is exothermic.

Student's answer

The rate of the reverse reaction increases because the forward reaction is exothermic.

Examiner's comment

The student should not have used the word 'rate'. If the temperature of an equilibrium system is increased, the rate of both forward and reverse reactions are increased, because increase in temperature increases the rate of all reactions (except enzyme catalysed reactions).

The correct answer is a statement that says

'The equilibrium shifts to the left in the endothermic direction'

or

'The equilibrium shifts to the left because the forward reaction is exothermic'

Students are advised to treat equilibrium and rate as two completely separate things which are not related to each other.

● Redox

Oxidation was originally defined as gain of oxygen. In the following reaction

$$2Mg_{(s)} + O_{2(g)} \rightarrow 2MgO_{(s)}$$

magnesium is oxidised because it gains oxygen. Because oxygen causes the oxidation, oxygen is the **oxidising agent**.

Reduction is the opposite of oxidation and was originally defined as loss of oxygen. In the following reaction

$$CuO_{(s)} + H_{2(g)} \rightarrow Cu_{(s)} + H_2O_{(l)}$$

copper(II) oxide is reduced because it loses oxygen. Hydrogen is the **reducing agent**. In this reaction, hydrogen gains oxygen, therefore hydrogen is oxidised. It follows that oxidation and reduction always occur at the same time.

A reaction in which oxidation and reduction both occur is known as a **redox reaction**.

Another example of a redox reaction involving oxygen is

$$4FeO_{(s)} \ + \ O_{2(g)} \ \rightarrow \ 2Fe_2O_{3(s)}$$
$$\text{iron(II) oxide} \qquad\qquad \text{iron(III) oxide}$$

in which iron(II) oxide is oxidised to iron(III) oxide using oxygen as an oxidising agent. This gives rise to another type of reaction in which an element is oxidised from a lower oxidation state to a higher oxidation state as in iron being oxidised from +2 to +3.

However, gain of oxygen and/or loss of hydrogen are very limited definitions of oxidation and reduction, because many redox reactions do not involve oxygen or hydrogen. The following reaction

$$2FeCl_{2(aq)} + Cl_{2(g)} \rightarrow 2FeCl_{3(aq)}$$
iron(II) chloride iron(III) chloride

in which iron(II) chloride is oxidised to iron(III) chloride using chlorine as the oxidising agent is an example.

The ionic equation for this reaction is

$$2Fe^{2+}_{(aq)} + Cl_{2(g)} \rightarrow 2Fe^{3+}_{(aq)} + 2Cl^-_{(aq)}$$

This can be broken down into two ionic half-equations:

Oxidation $2Fe^{2+}_{(aq)} \rightarrow 2Fe^{3+}_{(aq)} + 2e^-$

and

Reduction $2e^- + Cl_{2(g)} \rightarrow 2Cl^-_{(aq)}$

- Fe^{2+} is oxidised to Fe^{3+} by loss of electrons. Cl_2 is the oxidising agent.
- Therefore oxidation is electron loss.
- Oxidising agents are electron acceptors.
- Cl_2 is reduced to $2Cl^-$ by gain of electrons. Fe^{2+} is the reducing agent.
- Therefore reduction is electron gain.
- Reducing agents are electron donors.

In any redox reaction, electrons are transferred from the reducing agent to the oxidising agent. The reducing agent is oxidised and the oxidising agent is reduced.

Examiner's tip

In ionic half-equations

- electrons appear on the right for oxidation
- electrons appear on the left for reduction.

Testing for oxidising and reducing agents

Aqueous potassium manganate(VII) is an oxidising agent which can be used to test for the presence of reducing agents. When a reducing agent is added, the aqueous potassium manganate(VII) changes colour from purple to colourless.

Aqueous potassium iodide is a reducing agent which can be used to test for the presence of oxidising agents. When an oxidising agent is added, the aqueous potassium iodide changes colour from colourless to brown.

Exam-style questions

1 State whether the following changes are physical changes or chemical changes.

a Dissolving sodium chloride in water. [1 mark]

b The electrolysis of aqueous sodium chloride. [1 mark]

c Exposing a precipitate of silver chloride to sunlight. [1 mark]

d Fractional distillation of liquid air. [1 mark]

e Separating the dyes in ink by chromatography. [1 mark]
[Total: 5 marks]

2 When an excess of marble chips (calcium carbonate) is added to 50 cm³ of 0.10 mol dm⁻³ hydrochloric acid at 25 °C, the following reaction occurs:

$$CaCO_{3(s)} + 2HCl_{(aq)} \rightarrow CaCl_{2(aq)} + CO_{2(g)} + H_2O_{(l)}$$

The volume of carbon dioxide gas was collected in a gas syringe and measured at regular time intervals. This is experiment 1.

The experiment was repeated as shown in the table below. Graphs were plotted in each case (Figure 7.7).

The calcium carbonate is in excess in all five experiments.

Copy and complete the table below to show which graph corresponds to each different experiment. Each letter may be used once, more than once or not at all.

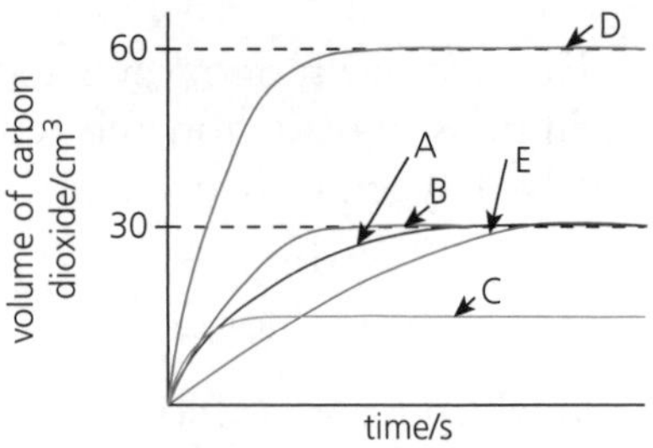

Figure 7.7 Graph showing volume of carbon dioxide collected

Experiment	Hydrochloric acid	Calcium carbonate	Temperature/°C	Graph	
1	50 cm³ of 0.10 mol dm⁻³	Marble chips	25	A	
2	**50 cm³ of 0.20 mol dm⁻³**	Marble chips	25		[1 mark]
3	50 cm³ of 0.10 mol dm⁻³	**Powdered**	25		[1 mark]
4	50 cm³ of 0.10 mol dm⁻³	Marble chips	**12.5**		[1 mark]
5	50 cm³ of 0.10 mol dm⁻³	Marble chips	**50**		[1 mark]

[Total: 4 marks]

3 State in which directions (if any) the following equilibrium mixtures would shift if the pressure on the system was increased. Explain your answer in each case.

a $H_{2(g)} + I_{2(g)} \rightleftharpoons 2HI_{(g)}$ [1 mark]

b $2SO_{2(g)} + O_{2(g)} \rightleftharpoons 2SO_{3(g)}$ [1 mark]

c $N_2O_{4(g)} \rightleftharpoons 2NO_{2(g)}$ [1 mark]

[Total: 3 marks]

4 State in which directions (if any) the following equilibrium mixtures would shift if the temperature on the system was decreased. Explain your answer in each case.

a $2SO_{2(g)} + O_{2(g)} \rightleftharpoons 2SO_{3(g)}$ exothermic in the forward direction [1 mark]

b $N_2O_{4(g)} \rightleftharpoons 2NO_{2(g)}$ endothermic in the forward direction [1 mark]

[Total: 2 marks]

5 The equation for the reaction between magnesium and copper(II) sulfate solution is shown.

$$Mg_{(s)} + CuSO_{4(aq)} \rightarrow MgSO_{4(aq)} + Cu_{(s)}$$

a Write an ionic equation for the reaction. [1 mark]

b Write two ionic half-equations representing oxidation and reduction in the reaction. [2 marks]

c State the formula of the species which acts as an oxidising agent in the reaction. Explain your answer. [2 marks]

d State the formula of the species which acts as a reducing agent in the reaction. Explain your answer. [2 marks]

[Total: 7 marks]

Acids, bases and salts

8

Key objectives

By the end of this section, you should be able to

- describe the characteristic properties of acids in their reactions with metals, bases and carbonates, and their effect on litmus and methyl orange
- describe the characteristic properties of bases in their reactions with acids and with ammonium salts and their effect on litmus and methyl orange
- describe neutrality, relative acidity and alkalinity in terms of pH measured using Universal indicator paper

- describe and explain the importance of controlling acidity in soil
- classify oxides as either acidic or basic related to metallic and non-metallic character
- suggest methods of preparation of salts
- define acids and bases in terms of proton transfer
- describe the meaning of weak and strong acids and bases
- classify other oxides as neutral or amphoteric.

● Key terms

Acid	A proton donor
Base	A proton acceptor
Strong acid	Exists completely as ions in aqueous solution
Weak acid	Only partially ionised in aqueous solution
Strong base	Exists completely as ions in aqueous solution
Salt	An ionic substance formed when the positive hydrogen ions in an acid are replaced by positive metallic ions or ammonium ions

● Acids

Acids are proton (H^+) donors.

The common laboratory **strong acids** are dilute hydrochloric, nitric and sulfuric acids. Their formulae are given below.

- Hydrochloric acid: HCl
- Nitric acid: HNO_3
- Sulfuric acid: H_2SO_4

In aqueous solution strong acids do not contain any particles with these formulae because they exist completely as ions, i.e.

$$HCl_{(aq)} \rightarrow H^+_{(aq)} + Cl^-_{(aq)}$$
$$HNO_{3(aq)} \rightarrow H^+_{(aq)} + NO_3^-_{(aq)}$$
$$H_2SO_{4(aq)} \rightarrow 2H^+_{(aq)} + SO_4^{2-}_{(aq)}$$

Weak acids, such as ethanoic acid, $CH_3COOH_{(aq)}$, exist mainly as covalent molecules with the formula $CH_3COOH_{(aq)}$, a small number of which dissociate into ions, i.e.

$$CH_3COOH_{(aq)} \rightleftharpoons CH_3COO^-_{(aq)} + H^+_{(aq)}$$

Reactions of acids

With metals

Acids react with **metals** above hydrogen in the reactivity series (although it would be dangerous to use a Group I metal or anything below calcium in Group II in a reaction with acids). The general equation is

acid + metal → salt + hydrogen

The metal dissolves, bubbles are seen and a solution of the **salt** forms whose colour depends on the metal used.

An example is

$$Zn_{(s)} + H_2SO_{4(aq)} \rightarrow ZnSO_{4(aq)} + H_{2(g)}$$

With carbonates

Acids react with both soluble and insoluble **carbonates**. The general equation is

acid + carbonate → salt + water + carbon dioxide

Solid carbonates dissolve, bubbles are seen and an aqueous solution (whose colour depends on the carbonate used) of the salt forms. An example is

$$CuCO_{3(s)} + 2HNO_{3(aq)} \rightarrow Cu(NO_3)_{2(aq)} + CO_{2(g)} + H_2O_{(l)}$$

This type of reaction occurs with carbonates either as solids or as aqueous solutions.

With bases

Acids react with all **bases** to form salts and water (in the case of ammonia, an ammonium salt is the only product).

The general equation is

acid + base → salt + water

With **insoluble bases** the solid dissolves and a solution forms. No bubbles are seen because no gas is produced.

An example is

$$Mg(OH)_{2(s)} + H_2SO_{4(aq)} \rightarrow MgSO_{4(aq)} + 2H_2O_{(l)}$$

With **alkalis** there are no observations as a colourless solution is produced from two colourless solutions.

An example is

$$2NaOH_{(aq)} + H_2SO_{4(aq)} \rightarrow Na_2SO_{4(aq)} + 2H_2O_{(l)}$$

With ammonia

The general equation is

acid + ammonia → ammonium salt

An example is

$$2NH_{3(aq)} + H_2SO_{4(aq)} \rightarrow (NH_4)_2SO_{4(aq)}$$

Strong and weak acids

Strong and weak acids can be distinguished experimentally by any of the following experimental methods.

	Strong acid	Weak acid
Add Universal indicator paper	Red or pH 0–2	Orange yellow or pH less than 7 and more than 2
Add magnesium ribbon	Bubbles quickly	Bubbles slowly
Add an insoluble carbonate, e.g. calcium carbonate	Bubbles quickly	Bubbles slowly
Set up a circuit with a bulb	Bulb lights brightly	Bulb lights dimly

Examiner's tip

● Many students have the impression that if an acid is weak, more of the weak acid is required to neutralise the same amount of alkali compared to a strong acid. The amount of any acid that is required to neutralise a given amount of alkali only depends on the number of moles of the acid and not whether the acid is strong or weak.

● Some questions begin with, 'How would you distinguish …?' For example, 'How would you distinguish between a strong acid and a weak acid?' The intention is that the student gives brief experimental details with results, for example, add magnesium ribbon to both and bubbles occur much faster with the strong acid than the weak acid. Instead students often answer with theory, for example, strong acids ionise completely and weak acids ionise partially. Although these statements are correct they do not address the question, 'How would you distinguish …?' It is also necessary to state what would happen with both substances or to give a comparison, rather than say the strong acid produces bubbles rapidly without reference to the weak acid.

● Bases

Bases are metallic oxides or hydroxides (or ammonia) which neutralise acids to form a salt and water.

● Bases that are soluble in water are called alkalis.
● Bases that do not dissolve in water are known as **insoluble bases**.

Alkalis are hydroxides or oxides of metals (or ammonia) that produce $OH^-_{(aq)}$ when dissolved in water.

The two laboratory strong alkalis are aqueous sodium hydroxide and potassium hydroxide. They both exist completely as ions in aqueous solution.

$$NaOH_{(aq)} \rightarrow Na^+_{(aq)} + OH^-_{(aq)}$$
$$KOH_{(aq)} \rightarrow K^+_{(aq)} + OH^-_{(aq)}$$

Ammonia solution is a weak base. An aqueous solution of ammonia exists mainly as NH_3 molecules, a small number of which react with water molecules to produce ions.

$$NH_{3(aq)} + H_2O_{(l)} \rightleftharpoons NH_4^+_{(aq)} + OH^-_{(aq)}$$

$NH_{3(aq)}$ accepts H^+ from $H_2O_{(l)}$ forming $NH_4^+_{(aq)}$ which shows that bases are **proton acceptors** by definition.

Reactions of bases

As described above, bases neutralise acids.

Application: Plants need soil to be at a specific pH to grow well. Soil acidity can be neutralised by the addition of a suitable base, such as calcium hydroxide, $Ca(OH)_{2(s)}$, also known as slaked lime.

With ammonium salts

Insoluble bases and alkalis react when heated with ammonium salts. Ammonia gas is given off.

The general equation is

    base + ammonium salt → salt + ammonia + water

An example is

$$Ca(OH)_{2(s)} + 2NH_4Cl_{(s)} \rightarrow CaCl_{2(s)} + 2NH_{3(g)} + 2H_2O_{(g)}$$

Examiner's tip

Many students do not apply the rules for writing formulae of ionic compounds when writing equations of this type. In this example, the formula of calcium chloride is often written as CaCl, without consideration of the charges on the ions present (see Chapter 3 for details).

● Indicators

Methyl orange and litmus can be used to indicate whether substances are acids or alkalis, but give no information about acid strength.

	Methyl orange	Litmus
Colour in acidic solution	Red	Red
Colour in neutral solution	Orange	Purple
Colour in alkaline solution	Yellow	Blue

The **pH scale** (Figure 8.1) uses numbers to distinguish between acids and alkalis of different strengths.

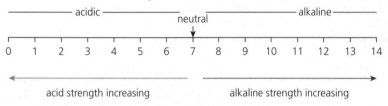

Figure 8.1 The pH scale

Aqueous solutions of acids have pH less than 7. Aqueous solutions of alkalis have pH more than 7. Neutral solutions have pH of 7.

The lower the pH numbers, the stronger the acid. The higher the pH numbers, the stronger the alkali.

Strong acids are regarded as having a pH of 0–2. Strong alkalis are regarded as having a pH of 12–14.

Universal indicator has different colours to show approximate pH numbers as shown.

Approximate pH	Colour of Universal indicator
Less than 3	Red
3–6	Orange–yellow
7	Green
8–11	Blue
More than 11	Purple

● Oxides

Oxides can be put into four categories.

1 **Acidic oxides** are non-metallic oxides that neutralise alkalis and form salts. Examples are carbon dioxide, CO_2, nitrogen dioxide, NO_2, and sulfur dioxide, SO_2. These oxides all dissolve in water and react with the water to form acids.

2 **Basic oxides** are metallic oxides that neutralise acids and form salts. Examples are magnesium oxide, MgO, calcium oxide, CaO, and copper(II) oxide, CuO. Some basic oxides dissolve in water to form alkaline hydroxides, whereas others are insoluble in water.

3 Some non-metallic oxides are **neutral oxides** which means that they do not react with either acids or alkalis. An example is carbon monoxide, CO.

4 Some metallic oxides are **amphoteric oxides** which means that they react with both acids and alkalis to form salts. Examples are zinc oxide, ZnO, and aluminium oxide, Al_2O_3.

Cambridge IGCSE Chemistry Study and Revision Guide © David Besser

Salts

Salts are ionic substances formed when the positive hydrogen ions in an acid are replaced by positive metallic ions or ammonium ions.

Salts can be made by different experimental methods, depending on their solubility in water.

Solubility of salts

Type of salt	Soluble		Insoluble
Nitrates	All nitrates are soluble in water		None
Sodium, potassium and ammonium salts	All sodium, potassium and ammonium salts are soluble in water		None
Chlorides	Chlorides are soluble in water except	→	Lead chloride and silver chloride are insoluble in water
Sulfates	Sulfates are soluble in water except	→	Lead sulfate and barium sulfate are insoluble in water
Lead salts	Lead nitrate is the only soluble lead salt	→	All other lead salts are insoluble
Carbonates	Only sodium, potassium and ammonium carbonates are soluble in water	→	All other carbonates are insoluble

Preparation of salts

There are three general methods of preparation of solid salts.

In all cases, details of crystallisation, washing and drying can be found in Chapter 2.

Method 1: Adding an excess of an insoluble base or insoluble carbonate or metal to a dilute acid.

Method 2: Titration using an acid and an alkali or a soluble carbonate.

Methods 1 and 2 can only be used for salts that are soluble in water.

Method 3: Mixing two solutions to obtain a salt that is insoluble in water by precipitation.

Examples

Method 1

Copper(II) sulfate crystals can be made by this method.

- Add solid copper(II) oxide or copper(II) hydroxide or copper(II) carbonate to dilute sulfuric acid in a beaker.

- Stir and/or heat the mixture.

- Add the solid until it will no longer dissolve which means all the acid has reacted and the solid is in excess (if copper(II) carbonate is used, there will be no further bubbling when all the acid has reacted).

- Filter off the excess solid.

- Using copper(II) oxide the equation is

$$H_2SO_{4(aq)} + CuO_{(s)} \rightarrow CuSO_{4(aq)} + H_2O_{(l)}$$

- Make pure crystals of copper(II) sulfate by crystallisation, washing and drying (see Chapter 2).

Examiner's tip

Using acids to prepare salts:

- **hydrochloric acid, HCl,** is used to prepare **chlorides**

- **nitric acid, HNO_3,** is used to prepare **nitrates**

- **sulfuric acid, H_2SO_4,** is used to prepare **sulfates** (or **hydrogen sulfates**)

- the positive ion in the salt comes from the insoluble base or insoluble carbonate or metal or alkali.

Cambridge IGCSE Chemistry Study and Revision Guide © David Besser

Method 2

Sodium sulfate crystals can be made by this method.

- Carry out sufficient titrations to find out the exact volume of dilute sulfuric acid in a burette that is required to neutralise a given pipette volume of aqueous sodium hydroxide. Use methyl orange as a suitable indicator.

- Repeat the process without using the indicator, but using the same volume of acid and alkali as used in the titration.

$$2NaOH_{(aq)} + H_2SO_{4(aq)} \rightarrow Na_2SO_{4(aq)} + 2H_2O_{(l)}$$

- Make pure crystals of sodium sulfate by crystallisation, washing and drying (see Chapter 2).

If twice the volume of the same dilute sulfuric acid is used, or half the volume of the same aqueous sodium hydroxide, sodium hydrogen sulfate (an acid salt) is the salt produced, according to the equation

$$NaOH_{(aq)} + H_2SO_{4(aq)} \rightarrow NaHSO_{4(aq)} + H_2O_{(l)}$$

Crystals of sodium hydrogen sulfate can be made from this solution in the same way.

Method 3

Lead sulfate can be made by this method.

- Because lead nitrate is the only soluble lead salt, lead nitrate solution must be used and mixed with any soluble sulfate, such as aqueous sodium sulfate (dilute sulfuric acid could also be used because it contains aqueous sulfate ions).

- The precipitate of lead sulfate must be removed by filtration and then washed with distilled water and dried in a low oven or on a warm windowsill.

    The equation is

$$Pb(NO_3)_{2(aq)} + Na_2SO_{4(aq)} \rightarrow PbSO_{4(s)} + 2NaNO_{3(aq)}$$

An ionic equation for any precipitation reaction shows the two aqueous ions on the left and the solid precipitate on the right in all cases. In this case

$$Pb^{2+}_{(aq)} + SO_4^{2-}_{(aq)} \rightarrow PbSO_{4(s)}$$

Exam-style questions

1   There are three general methods of preparation of solid salts.

Method 1: Adding an excess of an insoluble base or insoluble carbonate or metal to a dilute acid.

Method 2: Titration using an acid and an alkali or a soluble carbonate.

Methods 1 and 2 can only be used for salts that are soluble in water.

Method 3: Mixing two solutions to obtain a salt that is insoluble in water by precipitation.

For each of the following salt preparations, choose method 1, 2 or 3, name any additional reagent which is required and write the equation.

a   Cobalt(II) chloride starting with the insoluble compound cobalt(II) carbonate.   [3 marks]

b   The insoluble salt lead iodide, from aqueous lead nitrate.   [3 marks]

c   Potassium nitrate from aqueous potassium hydroxide.   [3 marks]
                                                              [Total: 9 marks]

2   Give full experimental details of how you would make pure dry crystals of magnesium sulfate starting with magnesium carbonate. You should include an equation in your answer.   [Total: 10 marks]

3   You are provided with a mixture of scandium oxide and copper(II) oxide which are both solids. Scandium oxide is an amphoteric oxide and copper(II) oxide is a basic oxide. Describe how you could obtain a sample of pure copper(II) oxide from the mixture. Both solids are insoluble in water.   [Total: 6 marks]

Cambridge IGCSE Chemistry Study and Revision Guide © David Besser

9 The Periodic Table

Key objectives

By the end of this section, you should be able to

- describe the Periodic Table as a means of classifying elements and its use to predict properties of elements
- describe the change from metallic to non-metallic character across a period
- describe some properties of the Group I elements lithium, sodium and potassium and predict the properties of other Group I elements
- describe some properties of the Group VII elements chlorine, bromine and iodine and predict the properties of other Group VII elements

- describe the noble gases in Group VIII or 0 as being unreactive, monatomic gases and explain this in terms of electronic structure
- be able to describe some properties of transition elements
- describe and explain the relationship between group number, number of outer shell electrons and metallic/non-metallic character
- identify trends in groups given information about the elements concerned
- know that transition elements have variable oxidation states.

Key terms

Periodic Table	Contains all the elements arranged in order of proton number
Groups	The vertical columns in the Periodic Table
Periods	The horizontal rows in the Periodic Table
Alkali metals	Group I. Metals which react with water to produce alkaline solutions (such as lithium, sodium and potassium)
Halogens	Group VII. Non-metallic, diatomic molecules
Noble gases	Group 0. Unreactive, monatomic, colourless gases
Transition elements	Metals found in the elongated section of the Periodic Table between Groups II and III (such as copper, iron and nickel)

Periodic Table

The **Periodic Table** contains the elements arranged in order of proton number (atomic number).

- The vertical columns of elements are called **groups**.
- The horizontal rows of elements are called **periods**.

Across Periods 2 and 3, there is a gradual change from metals on the left-hand side to non-metals on the right-hand side.

Common error

- Students often think that the elements are arranged in order of mass number or relative atomic mass. Most of the relative atomic masses of the elements do increase as proton numbers increase, but in some places the relative atomic mass decreases, e.g. argon to potassium.

Atoms of elements in the same group have the same number of electrons in the outer shell. The number of electrons in the outer shell determines the chemical properties of the element.

The number of shells present in an atom of an element is the same as the period number in which the element is found in the Periodic Table.

Potassium has proton number 19 and therefore its electron configuration is 2,8,8,1. There is one electron in the outer shell which means potassium is in Group **I** of the Periodic Table. Potassium has electrons in four shells, which means it is in Period 4.

It is illegal to photocopy this page

Cambridge IGCSE Chemistry Study and Revision Guide © David Besser

● Group properties

Group I

The Group I elements are known as the **alkali metals** because they react with water to produce alkaline solutions. The Group I elements are very reactive metals.

The Group I elements, in order of increasing proton number, are lithium, sodium, potassium, rubidium, caesium and francium. Only lithium, sodium and potassium are found in school laboratories, because rubidium, caesium and francium are dangerously reactive and francium is also radioactive.

The Group I elements

- are stored under oil because they react rapidly with oxygen in the air
- are good conductors of heat and electricity
- can be cut with a knife because they are soft
- are shiny when cut, but tarnish rapidly due to reaction with oxygen in the air
- have low densities and low melting points and boiling points compared to transition metals.

The melting points and boiling points of Group I elements decrease down the group. Densities change in an irregular manner.

Reaction with water

All Group I elements react vigorously with water at room temperature. The reactions are usually carried out in a glass trough. Observations are

- the metal moves around and floats on the surface of the water
- the reaction produces heat which causes the metal to melt as it reacts
- bubbles of hydrogen gas are given off
- the metal rapidly disappears, forming a colourless solution of the alkaline metal hydroxide.

The equation for the reaction with sodium is

$$2Na_{(s)} + 2H_2O_{(l)} \rightarrow 2NaOH_{(aq)} + H_{2(g)}$$

The equations with all the other Group I metals would be exactly the same (including balancing numbers) if the symbols for the other metals replaced that of Na in the above equation.

Reactivity of the Group I metals increases down the group. Sodium moves around the surface faster than lithium and also disappears more rapidly. Potassium bursts into a lilac flame. If rubidium and caesium are added to water an explosive reaction occurs, which is why they are not kept in school laboratories.

Group VII

The Group VII elements are known as the **halogens**.

The Group VII elements in order of increasing proton number are fluorine, chlorine, bromine, iodine and astatine. Only chlorine, bromine and iodine are found in school laboratories. Fluorine is too reactive to be used in schools and astatine is radioactive.

The elements are all non-metallic and exist as diatomic molecules (molecules containing two atoms). The appearance of those found in schools is shown in Table 9.1.

Examiner's tip

Examination questions often ask for observations, or ask, 'What would you see ...?' in a particular chemical or physical change. When Group I metals react with water, suitable observations are

- the metal disappears
- the metal melts
- bubbles/fizzing/effervescence (these all effectively mean the same thing)
- the metal floats and moves around on the surface
- potassium and those Group I metals below potassium burst into flames.

However, the following are not observations:

- names of the products
- a gas is given off (it is not possible to see a colourless gas)
- an alkaline solution forms (it is not possible to see that a solution is alkaline by observation alone)
- colour change of an indicator (unless an indicator is mentioned in the question).

Table 9.1 Physical appearance of chlorine, bromine and iodine

Element	Appearance at room temperature
Chlorine	Pale green gas
Bromine	Red-brown liquid
Iodine	Grey-black solid

The colours become darker as the group is descended. The change in physical state from gas to liquid to solid down the group indicates an increase in melting point and boiling point and density down the group (due to an increase in the strength of intermolecular forces).

Halogen displacement reactions

If halogens (or solutions of the halogens in water) are added to colourless aqueous solutions of potassium halides (chlorides, bromides and iodides), **displacement reactions** occur in the examples highlighted in Table 9.2.

Table 9.2 Displacement reactions observed in chlorine, bromine and iodine

	Aqueous potassium chloride, KCl	Aqueous potassium bromide, KBr	Aqueous potassium iodide, KI
Chlorine, Cl_2		Solution turns orange/yellow (bromine produced)	Solution turns brown (iodine produced)
Bromine, Br_2	No change		Solution turns brown (iodine produced)
Iodine, I_2	No change	No change	

An example of a displacement reaction is when chlorine displaces bromine from an aqueous solution of potassium bromide. The equation is

$$Cl_{2(g)} + 2KBr_{(aq)} \rightarrow 2KCl_{(aq)} + Br_{2(aq)}$$

As can be seen from Table 9.2,

- chlorine displaces bromine and iodine
- bromine displaces iodine, but does not displace chlorine
- iodine does not displace chlorine or bromine.

Halogens higher up the group can displace those lower down. Alternatively, we can say that reactivity increases up Group VII. This is opposite to the trend in reactivity shown in Group I.

We can use this information to make predictions about the other halogens and halides (see the Exam-style questions at the end of the chapter).

Group 0

The Group 0 elements are known as the **noble gases**. The Group 0 elements, in order of increasing proton number, are helium, neon, argon, krypton, xenon and radon.

The Group 0 elements

- are all colourless gases
- are all monatomic; this means they exist as individual atoms because their atoms all have a full outer shell of electrons and do not form covalent bonds with other Group 0 atoms to form diatomic molecules
- are very unreactive because their atoms all have a full outer shell of electrons, therefore they do not need to share, lose or gain electrons to achieve a full outer shell of electrons.

Uses of Group 0 elements

- Helium is used in filling balloons because of its low density. Its unreactive nature means that hazards associated with hydrogen are absent if helium is used.
- Argon is used in light bulbs to prevent the tungsten filament from burning. This is because argon does not support burning due to its unreactivity.

● Transition elements

Transition elements are all metals (also known as transition metals). They are found in the elongated section of the Periodic Table between Groups II and III. Common examples are copper, iron and nickel.

Physical properties

Transition elements are all metals and show the usual physical properties of metals (see Chapter 10). In addition, transition metals

- are harder and stronger than the elements in Group I
- have higher densities than the elements in Group I
- have higher melting points than the elements in Group I.

Chemical properties

- Transition elements form coloured compounds, e.g. copper(II) sulfate crystals are blue and potassium manganate(VII) is purple.
- The elements and their compounds show catalytic activity, e.g. iron in the Haber process and vanadium(V) oxide in the Contact process.
- Transition elements have variable oxidation states, e.g. iron can form Fe^{2+} and Fe^{3+} ions.

Exam-style questions

1 The diagram below shows part of the Periodic Table. Use the letters A to H inclusive to answer the questions that follow. Each letter may be used once, more than once, or not at all. Give the letter that represents

   a the Group I element that is most reactive   [1 mark]

   b the Group VII element that is most reactive   [1 mark]

   c a transition element   [1 mark]

   d an element in Period 3   [1 mark]

   e an element whose atoms have four electrons in their outer shell.   [1 mark]
                                                             [Total: 5 marks]

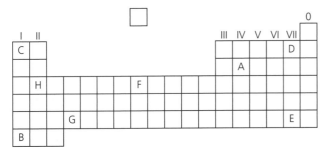

Figure 9.1

Cambridge IGCSE Chemistry Study and Revision Guide © David Besser

2   Lithium is added to cold water in a glass trough.

    a   Give three observations you would expect to make.   [3 marks]

    b   Write a chemical equation for the reaction that occurs. Include
       state symbols.   [3 marks]

    c   If methyl orange is added to the liquid in the trough after the
       reaction, what colour would it turn to?   [1 mark]
                                     [Total: 7 marks]

3   Use the table of halogen displacement reactions to answer the following
    questions. Write chemical equations (with state symbols) and ionic
    equations for the reactions that occur between

    a   chlorine and aqueous potassium iodide   [4 marks]

    b   bromine and aqueous potassium iodide.   [4 marks]
                                       [Total: 8 marks]

4   Use the Periodic Table to predict reactions that would occur between

    a   fluorine and aqueous potassium chloride

    b   astatine and aqueous potassium fluoride

    c   bromine and aqueous potassium astatide

    d   iodine and aqueous potassium fluoride.

    If you predict that a reaction would occur, write a chemical
    equation for the reaction. If you predict that a reaction
    would not occur write no reaction.   [Total: 6 marks]

5   Copper and iron have variable oxidation states. State the formulae of

    a   copper(I) oxide   [1 mark]

    b   copper(II) nitrate   [1 mark]

    c   iron(II) chloride   [1 mark]

    d   iron(III) sulfate.   [1 mark]
                                       [Total: 4 marks]

(10) Metals

Key objectives

By the end of this section, you should be able to

- describe the physical properties of metals
- describe the chemical properties of metals
- identify representations of alloys from diagrams of structure and explain in terms of properties why alloys are used instead of pure metals
- place the following in order of reactivity: potassium, sodium, calcium, magnesium, zinc, iron, (hydrogen) and copper by reference to the reactions with
  - water or steam
  - dilute hydrochloric acid and
  - the reduction of their oxides with carbon
- deduce an order of reactivity from a given set of experimental results
- describe the ease of obtaining metals from their ores by reference to the position of elements in the reactivity series
- describe and state the essential reactions in the extraction of iron from hematite
- describe the conversion of iron into steel using basic oxides and oxygen

- know that aluminium is extracted from its ore bauxite by electrolysis
- state the advantages and disadvantages of recycling iron/steel and aluminium
- name some uses of aluminium
- name some uses of copper
- name some uses of mild steel and stainless steel
- describe the reactivity series as related to the tendency of a metal to form its positive ion, illustrated by reaction with
  - the aqueous ions
  - the oxides of the other listed metals
- account for the apparent unreactivity of aluminium in terms of the oxide layer which adheres to the metal
- describe and explain the action of heat on the hydroxides, carbonates and nitrates of the listed metals
- describe in outline the extraction of zinc from zinc blende
- describe in outline the extraction of aluminium from bauxite, including the role of cryolite and the reactions at the electrodes
- explain the uses of zinc in galvanising and for making brass.

● Properties of metals

Physical properties

The physical properties of metals are shown in Table 10.1 (see also Chapter 9).

Table 10.1 Physical properties of metals

Property	Metallic property
Physical state at room temperature	Solid (except mercury)
Malleability	Good
Ductility	Good
Appearance	Shiny (lustrous)
Melting point and boiling point	Usually high
Density	Usually high
Conductivity (thermal and electrical)	Good

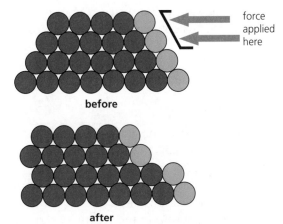

Figure 10.1 The positions of the positive ions in a metal before and after a force has been applied

Metals are **malleable** (can be hammered into different shapes) and **ductile** (can be drawn into wires). Although metallic bonds are strong, they are not rigid, which means that the rows of ions in metals can slide over each other when a force is applied (Figure 10.1). This is because the ions in a metallic element are all the same size.

When a metallic object is required to be particularly strong, an **alloy** is often used instead of a metallic element. In alloys such as brass, bronze and steel, the metallic element is mixed with small amounts of another element or elements. The ions or atoms of the other elements are a different size to those of the main element. This prevents the rows of metallic ions from sliding over each other. Therefore, an alloy retains its shape much better than a pure metal when a force is applied.

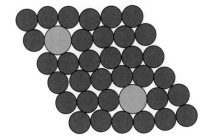

Figure 10.2 Alloy structure

It is illegal to photocopy this page

Cambridge IGCSE Chemistry Study and Revision Guide © David Besser

● Reactivity series

Metals can be placed in order of their reactivity with other elements. This is known as the **reactivity series** (see right).

The positions of carbon and hydrogen are inserted for reference to displacement reactions. Potassium, sodium and calcium react with cold water to produce an aqueous solution of an alkaline hydroxide and hydrogen gas (see Chapter 9), for example

$$Ca_{(s)} + 2H_2O_{(l)} \rightarrow Ca(OH)_{2(aq)} + H_{2(g)}$$

Magnesium, zinc and iron react extremely slowly with cold water. They do react more rapidly when heated with steam, for example

$$Mg_{(s)} + H_2O_{(g)} \rightarrow MgO_{(s)} + H_{2(g)}$$

Metals above hydrogen in the reactivity series react with dilute acids to produce a salt and hydrogen (the reaction of potassium and sodium with acids would be too dangerous to carry out in school laboratories).

$$Fe_{(s)} + 2HCl_{(aq)} \rightarrow FeCl_{2(aq)} + H_{2(g)}$$

Metals below hydrogen in the reactivity series, e.g. copper, do not react with cold water, steam or dilute acids.

Oxides of metals below carbon in the reactivity series are reduced to the metal when heated with carbon, for example

$$2CuO_{(s)} + C_{(s)} \rightarrow 2Cu_{(s)} + CO_{2(g)}$$

MOST REACTIVE
potassium
↓
sodium
↓
calcium
↓
magnesium
↓
aluminium
↓
(carbon)
↓
zinc
↓
iron
↓
(hydrogen)
↓
copper
LEAST REACTIVE

Displacement reactions

Metals will displace other metals from aqueous solutions of their ions, e.g. magnesium will displace copper from an aqueous solution containing its ions such as copper(II) sulfate solution.

$$Mg_{(s)} + CuSO_{4(aq)} \rightarrow Cu_{(s)} + MgSO_{4(aq)}$$

The ionic equation is

$$Mg_{(s)} + Cu^{2+}_{(aq)} \rightarrow Cu_{(s)} + Mg^{2+}_{(aq)}$$

The reaction occurs because magnesium is a more reactive metal than copper. This means that magnesium has a greater tendency to form positive ions than copper.

If copper was added to a solution containing magnesium ions, such as aqueous magnesium sulfate, no reaction occurs.

Metals higher up in the reactivity series will displace those lower down.

A similar reaction occurs if a metal oxide is heated with a more reactive metal. For example, if zinc powder is heated with copper(II) oxide, the following reaction occurs:

$$CuO_{(s)} + Zn_{(s)} \rightarrow Cu_{(s)} + ZnO_{(s)}$$

However, if zinc oxide is heated with copper, no reaction occurs.

Cambridge IGCSE Chemistry Study and Revision Guide © David Besser

Deducing order of reactivity of metals

To put metals in order of reactivity, reactions can be attempted as in, for example, adding a metal to an aqueous solution containing ions of another metal or heating a metal with the oxide of another metal.

Another method involves the use of electrochemical cells (see Chapter 5) containing two dissimilar metals in an electrolyte (Figure 10.3).

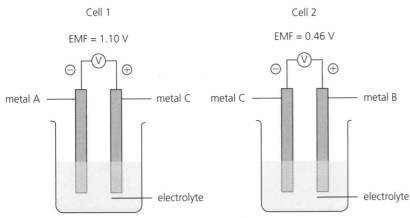

Figure 10.3 Deducing order of reactivity using chemical cells

In electrochemical cells of this type, the more reactive metal is the negative terminal, because the more reactive metal is the one with the greater tendency to release electrons and form positive ions. The reading on the voltmeter (known as the cell EMF (electromotive force)) represents the difference in reactivity between the two metals.

Cell 1 tells us that metal A (–) is higher in the reactivity series than metal C (+) and 1.10 V is a measurement of the difference in reactivity.

Cell 2 tells us that metal C (–) is higher in the reactivity series than metal B (+) and 0.46 V is a measurement of the difference in reactivity.

Therefore, the order of the three metals in the reactivity series is

Most reactive: A → C → B: Least reactive

Examiner's tip

If a cell was set up with A and B in an electrolyte

● A would be the negative electrode

● B would be the positive electrode

● the cell EMF would be 1.10 V + 0.46 V = 1.56 V.

Unexpected behaviour of aluminium

Aluminium appears between magnesium and zinc in the reactivity series. However, aluminium often appears to be much less reactive than its position in the reactivity series suggests.

For example, if aluminium is placed in an aqueous solution of copper(II) sulfate, there is hardly any reaction until the layer of aluminium oxide is removed. This is because aluminium is so reactive that it reacts with the oxygen in the air forming a layer of aluminium oxide which adheres to the aluminium underneath and protects the metal. Such a layer can be deliberately placed onto the surface of aluminium metal by a process called **anodising**. This means that aluminium can be used for things which would not normally be associated with reactive metals, such as aeroplane bodies, cooking foil and pots and pans.

Cambridge IGCSE Chemistry Study and Revision Guide © David Besser

Thermal decomposition of metallic compounds

The ease with which metallic hydroxides, nitrates and carbonates decompose when heated is related to the position of the metallic element in the reactivity series.

Compounds of metals at the top of the reactivity series either require a large amount of heat energy to decompose or they do not decompose at all. We say that they are stable to heat. As the reactivity of the metallic elements decreases, their compounds are less stable to heat and are more easily decomposed by heat.

In the case of nitrates, Group I nitrates decompose partially, whereas the nitrates of the other listed metals decompose more completely.

Hydroxides

Hydroxides of very reactive metals do not decompose when heated. Those of less reactive metals decompose into their oxides and give off steam, for example,

$$Cu(OH)_{2(s)} \rightarrow CuO_{(s)} + H_2O_{(g)}$$

Nitrates

All nitrates decompose when heated, but nitrates do not all decompose to produce similar products.

Group I nitrates (except lithium nitrate) decompose partially to form the metallic nitrite and oxygen gas, for example,

$$2NaNO_{3(s)} \rightarrow 2NaNO_{2(s)} + O_{2(g)}$$

Other metallic nitrates of less reactive metals decompose more completely, producing the metal oxide and giving off nitrogen dioxide, a brown gas, and oxygen gas.

$$2Mg(NO_3)_{2(s)} \rightarrow 2MgO_{(s)} + 4NO_{2(g)} + O_{2(g)}$$

Carbonates

Group I carbonates (except lithium carbonate) do not decompose when heated. The carbonates of all the other listed metals decompose into the oxide and carbon dioxide gas. The amount of heat required for decomposition is greater for carbonates of very reactive metals, such as calcium carbonate,

$$CaCO_{3(s)} \rightarrow CaO_{(s)} + CO_{2(g)}$$

but carbonates of less reactive metals like copper(II) carbonate decompose at much lower temperatures

$$CuCO_{3(s)} \rightarrow CuO_{(s)} + CO_{2(g)}$$

Common error

- Students often state that metals higher up in the reactivity series decompose with much more difficulty than those lower down. It is not the metals that decompose; it is their compounds, namely the hydroxides, nitrates and carbonates. Metals, being elements, cannot decompose because an element is something that cannot decompose into anything simpler.

● Extraction of metals

Metals can be extracted from their ores more easily as we go down the reactivity series.

There are three general methods of extracting metals from their ores:

1 Metals of average reactivity, e.g. iron and zinc, are extracted by chemical reduction using carbon/carbon monoxide as reducing agents.

2 Metals of low reactivity, e.g. copper, are extracted by

  a chemical reduction using carbon/carbon monoxide as reducing agents or

  b electrolysis of aqueous solutions containing their ions.

3 Very reactive metals, e.g. potassium, sodium, calcium, magnesium and aluminium

  a cannot be extracted by chemical reduction because the ores are not reduced by chemical reducing agents, such as carbon, carbon monoxide or hydrogen

  b cannot be extracted by electrolysis of aqueous solutions, because hydrogen is formed at the cathode instead of the metal (see Chapter 5).

  Therefore, these metals are extracted by electrolysis of molten ionic compounds.

Aluminium

Aluminium is extracted from **bauxite** which is impure aluminium oxide, Al_2O_3. Bauxite is first purified and then electrolysis is carried out. Aluminium oxide is not reduced by carbon monoxide or any other common reducing agent, which means electrolysis has to be used. This is expensive due to the high cost of electricity.

Aluminium oxide has a melting point of 2017 °C which would require a large amount of heat energy to achieve and therefore would further increase costs. Instead, the aluminium oxide is dissolved in another aluminium compound, molten cryolite, $Na3AlF_6$. The advantages of dissolving aluminium oxide in molten cryolite are

● the electrolyte can be maintained in the liquid state between 800 °C and 1000 °C, a temperature considerably lower than 2017 °C, which greatly reduces energy costs

● cryolite improves the conductivity of the electrolyte.

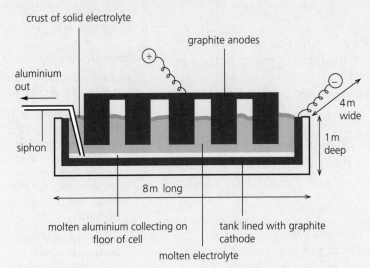

Figure 10.4 The Hall–Heroult cell is used in industry to extract aluminium

Cambridge IGCSE Chemistry Study and Revision Guide © David Besser

Aluminium oxide in molten cryolite behaves in the same way as molten aluminium oxide as far as the products of electrolysis are concerned.

Electrolysis is carried out in a steel tank using carbon (graphite) electrodes. The anodes are carbon (graphite) blocks that are lowered into the electrolyte. The cathode is the carbon (graphite) lining of the tank.

The electrode reactions are

$$\text{Cathode (−)} \quad Al^{3+}_{(l)} + 3e^- \rightarrow Al$$

$$\text{Anode (+)} \quad 2O^{2-}_{(g)} \rightarrow O_{2(g)} + 4e^-$$

Molten aluminium collects at the bottom of the tank and is siphoned off.

The oxygen that is produced at the anode reacts, at the high temperature of the cell, with the graphite anodes

$$C_{(s)} + O_{2(g)} \rightarrow CO_{2(g)}$$

producing carbon dioxide gas which escapes. Thus the anodes burn away and have to be replaced regularly.

The high cost of electricity is the largest expense for this process, which is carried out in regions where cheap electricity, e.g. from hydro-electric power, is available.

Iron

Iron is extracted from **hematite**, impure iron(III) oxide, Fe_2O_3, in a blast furnace.

Hematite, coke, C, and limestone, $CaCO_3$, are fed into the top of the blast furnace.

A blast of hot air enters near the bottom.

Coke reacts with the oxygen in the air forming carbon dioxide. The reaction is highly exothermic and provides the high temperature required for the other reactions.

$$C_{(s)} + O_{2(g)} \rightarrow CO_{2(g)}$$

The carbon dioxide reacts with more coke higher up to produce carbon monoxide in an endothermic reaction.

$$CO_{2(g)} + C_{(s)} \rightarrow 2CO_{(g)}$$

The carbon monoxide reduces the hematite to molten iron

$$Fe_2O_{3(s)} + 3CO_{(g)} \rightarrow 2Fe_{(l)} + 3CO_{2(g)}$$

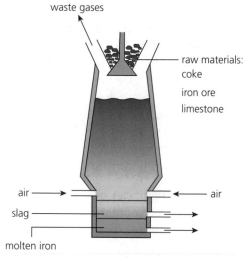

Figure 10.5 A blast furnace

and the molten iron trickles to the bottom and is tapped off.

The function of the limestone is to remove the main impurity in the iron ore which is silicon dioxide (silicon(IV) oxide).

The limestone decomposes at the high temperature inside the blast furnace.

$$CaCO_{3(s)} \rightarrow CaO_{(s)} + CO_{2(g)}$$

Calcium oxide then reacts with silicon(IV) oxide to form calcium silicate which forms a molten slag as a separate layer above the molten iron (it is less dense than iron).

$$CaO_{(s)} + SiO_{2(s)} \rightarrow CaSiO_{3(l)}$$

Slag is used by builders and road makers for foundations.

The iron produced in the blast furnace is called pig iron or cast iron. It contains about 4% carbon and its use is limited because it is brittle. The majority of pig iron is converted into steel.

Examiner's tip

You should know that the function of the coke is

● to act as a fuel, because the highly exothermic reaction with oxygen provides a high enough temperature for the reduction of the hematite

● to produce carbon monoxide which reduces the hematite to iron.

Conversion of iron into steel

Pig iron contains about 4% carbon and other non-metallic impurities, such as phosphorus, silicon and sulfur. The production of steel involves

- removing most of the non-metallic impurities (except small amounts of carbon)

- adding small, controlled amounts of transition elements (additives). The transition elements that are used and the proportions in which they are added determine the properties of the steels that are formed.

The impurities are removed by the basic oxygen process:

- Oxygen at high pressure is blown onto the surface of the molten metal.

- This causes oxidation of some of the carbon to carbon dioxide, and sulfur to sulfur dioxide, both of which escape as gases.

$$C_{(s)} + O_{2(g)} \rightarrow CO_{2(g)}$$
$$S_{(s)} + O_{2(g)} \rightarrow SO_{2(g)}$$

- Silicon is oxidised to silicon(IV) oxide and phosphorus is oxidised to phosphorus(V) oxide, which are both solids.

$$Si_{(s)} + O_{2(g)} \rightarrow SiO_{2(s)}$$
$$4P_{(s)} + 5O_{2(g)} \rightarrow P_4O_{10(s)}$$

- Calcium oxide (quicklime) is added. This reacts with the solid oxides to produce calcium silicate and calcium phosphate which are removed as molten slag.

$$CaO_{(s)} + SiO_{2(s)} \rightarrow CaSiO_{3(l)}$$
$$6CaO_{(s)} + P_4O_{10(s)} \rightarrow 2Ca_3(PO_4)_{2(l)}$$

- The amount of carbon remaining can be controlled depending on the type of steel required. Mild steel contains approximately 0.5% carbon and 99.5% iron.
- Transition elements are then added in exact quantities to produce different types of steel. Stainless steel contains 18% chromium, 8% nickel, as well as 74% iron.

Zinc

Zinc is extracted from zinc blende (impure zinc sulfide, ZnS) by a process similar to the blast furnace process, using carbon as the reducing agent.
   Zinc blende is purified. Zinc sulfide is then heated very strongly (roasted) in a current of air to convert it into zinc oxide.

$$2ZnS_{(s)} + 3O_{2(g)} \rightarrow 2ZnO_{(s)} + 2SO_{2(g)}$$

The zinc oxide is heated very strongly with powdered coke in a furnace.

$$ZnO_{(s)} + C_{(s)} \rightarrow Zn_{(g)} + CO_{(g)}$$

The zinc vapour then cools and condenses. It is removed as molten zinc.

Copper

The refining of copper is based on the electrolysis of aqueous copper(II) sulfate using copper electrodes (see Chapter 5).
   Refining means purification. Small amounts of impurities in copper cut down its electrical conductivity noticeably.

The electrolysis is carried out using impure copper as the anode (+), pure copper as the cathode (−) and aqueous copper(II) sulfate as the electrolyte.

The copper from the anode (+) goes into solution as positive ions.

$$Cu_{(s)} \rightarrow Cu^{2+}_{(aq)} + 2e^-$$

The impurities either go into solution as positive ions or fall off the anode and deposit at the bottom of the container. They are removed from time to time.

Copper metal forms at the cathode (−)

$$Cu^{2+}_{(aq)} + 2e^- \rightarrow Cu_{(s)}$$

Thus pure copper is formed on the cathode (which was originally made of pure copper). The cathode is removed and replaced from time to time.

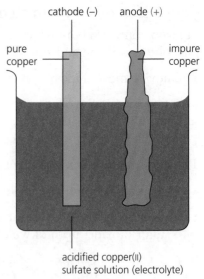

Figure 10.6 Copper purification process

Recycling aluminium and iron/steel

Much aluminium is recycled from drinks cans. Iron and steel are recycled from household goods.

The advantages of recycling are

- Natural resources of bauxite and hematite will last longer.
- Drinks cans and other household objects will not take up large amounts of room in landfill sites, where they react with oxygen from the air (thus removing it) as they are oxidised.
- The cost of recycling is much lower than extracting the metals from their ores, largely due to reduced energy costs.

The disadvantages of recycling are

- The collection and sorting of domestic materials to be recycled can be expensive, time consuming and require energy.
- The purity of metals obtained by recycling may not be as high as that obtained by extraction from metal ores.

● Uses of metals

Uses of some metals, related to their properties, are shown in Table 10.2.

Table 10.2 Uses of some metals

Metal	Use	Properties
Aluminium	Manufacture of aircraft	Strength and low density
	Food containers	Resistance to corrosion
Mild steel	Car bodies	High malleability, high tensile strength
	Machinery	
Stainless steel	Cutlery	Tough, does not corrode
	Chemical plants	
Copper	Electrical wiring	High electrical conductivity
	Cooking utensils	Appearance, high thermal conductivity, high melting point
Zinc	Galvanising	Protects iron from rusting
	Making brass	Brass is more ductile and stronger than copper

Cambridge IGCSE Chemistry Study and Revision Guide © David Besser

Exam-style questions

1 The results of some experiments carried out by adding a metal to aqueous solutions containing ions of another metal are shown in the table below, where ✓ is displacement reaction occurs and ✗ is no reaction occurs.

	A(NO$_3$)$_{2(aq)}$	B(NO$_3$)$_{2(aq)}$	C(NO$_3$)$_{2(aq)}$	D(NO$_3$)$_{2(aq)}$
Metal A$_{(s)}$	✗	✗	✗	✓
Metal B$_{(s)}$	✓	✗	✓	✓
Metal C$_{(s)}$	✓	✗	✗	✓
Metal D$_{(s)}$	✗	✗	✗	✗

  a Put the four metals in order of reactivity, starting with the most reactive first.   [1 mark]

  b Write a chemical equation for the reaction occurring when metal B is added to A(NO$_3$)$_{2(aq)}$.   [1 mark]

  c Write an ionic equation for the reaction occurring when metal C is added to D(NO$_3$)$_{2(aq)}$.   [1 mark]

  d If metals A and B are both placed in an aqueous solution of an electrolyte, which metal would be the negative electrode? Explain your answer.   [1 mark]

  e Write an equation for the reaction occurring when metal B is added to the oxide of metal D and the mixture is heated.   [1 mark]
[Total: 5 marks]

2 You are provided with a mixture of powdered copper and powdered zinc. Describe how you would obtain a sample of pure copper from the mixture. You should give all observations for any reactions that you describe.

   Note: neither metal dissolves in water.   [Total: 4 marks]

3 Impure nickel can be refined using a method similar to the method used to refine copper. Draw a diagram of the apparatus that you would use for the refining of nickel. You should fully label all the chemical substances in the diagram.   [Total: 4 marks]

4 Copper(ii) oxide → add dilute nitric acid → blue solution B → crystallisation → blue crystals B → heat crystals strongly → black solid C + brown gas D + colourless gas E

  a Name B, C, D and E.   [4 marks]

  b Write equations for the reactions that occur when

     i   copper(ii) oxide is added to dilute nitric acid

     ii blue crystals B are heated strongly.   [4 marks]

  c When copper(ii) oxide is reacted with dilute nitric acid, blue solution B is produced. Give the names of two other substances that could be used instead of copper(ii) oxide to produce blue solution B when reacted with dilute nitric acid.   [2 marks]
[Total: 10 marks]

11 Air and water

Key objectives

By the end of this section, you should be able to

- describe the chemical tests for water using cobalt(II) chloride and copper(II) sulfate
- describe in outline the treatment of the drinking water supply in terms of filtration and chlorination
- name some uses of water in industry and in the home
- state the composition of clean dry air as being approximately 78% nitrogen, 21% oxygen and the remainder being a mixture of noble gases and carbon dioxide
- name the common pollutants in the air as being carbon monoxide, sulfur dioxide, oxides of nitrogen and lead compounds and to state the sources of these pollutants
- state the adverse effects of these common pollutants and discuss why these pollutants are of global concern
- state the conditions required for the rusting of iron
- describe and explain coating methods of rust prevention

- describe the need for nitrogen, phosphorus and potassium in fertilisers
- describe the displacement of ammonia from ammonium salts (see Chapter 8)
- describe the separation of oxygen and nitrogen from liquid air by fractional distillation
- describe the catalytic removal of oxides of nitrogen from car exhaust gases
- describe and explain galvanising and sacrificial protection in terms of the reactivity series (see Chapter 10) as a method of rust prevention
- describe and explain the essential conditions for the Haber process for the manufacture of ammonia including the sources of hydrogen (from hydrocarbons or steam) and nitrogen (from air) (see also Chapter 7 on rates and equilibrium and Chapter 14 on cracking).

● Key terms

Filtration	A treatment for drinking water that involves passing impure water through screens to filter out floating debris
Chlorination	A treatment for drinking water in which small amounts of chlorine are added to kill bacteria
Fractional distillation	A method for separating the components of air
Catalytic converters	Present in car exhausts in some countries to remove pollutant gases and convert them into non-pollutant gases
Rust	Iron forms rust when it is exposed to oxygen and water. Rust is hydrated iron(III) oxide
Fertilisers	Substances added to the soil to supply nutrients that are essential for the growth of plants

● Water

Test for water

Tests for the presence of water can be carried out using anhydrous cobalt chloride or anhydrous copper(II) sulfate. The colour changes are shown in Table 11.1.

Table 11.1

	Original colour	Final colour
Anhydrous cobalt chloride	Blue	Pink
Anhydrous copper(II) sulfate	White	Blue

Cobalt chloride paper is commonly used instead of the anhydrous solid.

> **Examiner's tip**
>
> These substances are not used as a test for pure water. The colour changes shown occur if water or anything containing water (including all aqueous solutions) is used.
>
> Purity of water can be determined by measuring the boiling point, which is 100 °C at 1 atmosphere pressure.

It is illegal to photocopy this page

Cambridge IGCSE Chemistry Study and Revision Guide © David Besser

Treatment of drinking water

Impure water is made fit for drinking by the following methods:

- **Filtration**: This involves passing impure water through screens to filter out floating debris.
- **Chlorination**: Small amounts of chlorine gas are added to kill bacteria.

Common error

- Students often state that chlorine is added to purify water. Water containing small amounts of added chlorine should not be described as pure.

Uses of water

Water is used in industry

- as a solvent
- as a coolant
- for cleaning
- as a chemical reactant (e.g. in the hydration of ethene; see Chapter 15).

Water is used in the home

- in cooking
- in cleaning
- for drinking.

Air

Air is a mixture, and in common with all mixtures its composition can vary, particularly in industrial and rural areas.

The approximate composition of clean dry air is

- 78% nitrogen
- 21% oxygen
- 0.03% carbon dioxide
- 1% argon.

Very small amounts of other noble gases are also present.

Fractional distillation of liquid air

In order to separate the components of air,

- it is cooled to remove carbon dioxide and water vapour as solids
- it is compressed and expanded continually to liquefy it at $-200\,°C$
- the liquid air is then separated by **fractional distillation** into liquid oxygen, liquid nitrogen as well as liquid Group 0 elements.

Air pollution

Common gaseous pollutants, their sources and related pollution problems are shown in Table 11.2.

Table 11.2

Pollutant	Source	Pollution problems
Carbon monoxide, CO	Incomplete combustion of fossil fuels	Toxic
Sulfur dioxide, SO_2	Combustion of fossil fuels containing small amounts of sulfur	Irritation of the respiratory system Dissolves in rain water forming acid rain which causes damage to buildings made of, e.g. limestone and marble
Oxides of nitrogen (often represented as NO_x to signify more than one oxide of nitrogen) including nitrogen dioxide, NO_2	Nitrogen and oxygen (both from the air) react together at very high temperatures in car engines	Production of low-level ozone (respiratory system irritant) Nitrogen dioxide, NO_2, dissolves in rain water forming acid rain which causes damage to buildings made of, for example, limestone and marble Photochemical smog
Lead compounds	Lead compounds in petrol (only in countries where leaded petrol is still used)	Toxic

Catalytic converters

Catalytic converters are present in car exhausts in some countries. Their purpose is to remove pollutant gases and convert them into non-pollutant gases.

The catalysts present in catalytic converters include alloys containing transition elements, such as platinum, rhodium and palladium. There are several reactions occurring inside catalytic converters including

$$2CO_{(g)} + O_{2(g)} \rightarrow 2CO_{2(g)}$$

and

$$2CO_{(g)} + 2NO_{(g)} \rightarrow N_{2(g)} + 2CO_{2(g)}$$

Thus pollutant gases are converted into non-pollutant gases.

Common errors

There are probably more misunderstandings concerning atmospheric pollution than any other Chapter.

Many students think that all atmospheric pollutants are responsible for all environmental problems, particularly global warming. It is very important that students study Table 11.2 carefully and learn about the sources and pollution problems caused by each individual pollutant.

The following points need to in be noted:

- Oxides of nitrogen are produced by reaction between nitrogen and oxygen, both of which come from the air. The nitrogen is not present in the fuel.

- Oxides of nitrogen are often mistakenly thought to be produced in car exhausts as opposed to in car engines.

- Pollutants are removed by catalytic converters in car exhausts.

- If cars do not have efficient catalytic converters, the gaseous pollutants enter the atmosphere through the car exhaust.

- Sulfur dioxide is often regarded as being produced by the deliberate burning of sulfur, as opposed to the small amounts of sulfur impurities in fossil fuels.

Rusting of iron

Rust can be described as hydrated iron(III) oxide with a formula that can be represented as $Fe_2O_3.xH_2O$ (*x* is used because the amount of water of crystallisation varies from one sample of rust to another).

Iron only forms rust when it is exposed to oxygen (e.g. from the air) and water.

Prevention of rusting

Rusting can be prevented by coating the iron with

- paint
- oil or grease
- plastic
- other metals, such as zinc (which is known as **galvanising**).

These methods all work by preventing oxygen and water coming into contact with the iron and thus preventing a reaction taking place.

Sacrificial protection

Some metals will continue to prevent iron from rusting even if the surface is scratched. Such metals must be above iron in the reactivity series, but must not be so reactive that they will react rapidly with water themselves. Zinc and magnesium are both used in this way. Tin, which is below iron in the reactivity series, will only protect iron if it is not scratched (Figure 11.1).

In addition, bars of zinc may be attached to the hulls of ships without attempting to cover the surface of the iron. Rusting will not occur in these circumstances (Figure 11.2).

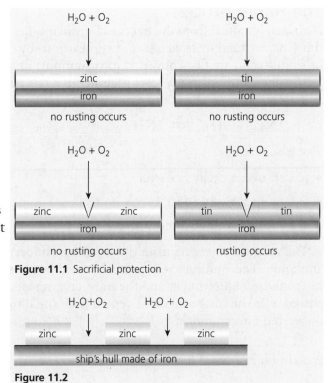

Figure 11.1 Sacrificial protection

Figure 11.2

The first stage of rusting is the oxidation of iron to iron(II) ions by oxygen in the presence of water:

$$Fe \rightarrow Fe^{2+} + 2e^-$$

If zinc is present, even if the zinc is scratched, the zinc will be oxidised in preference to the iron.

$$Zn \rightarrow Zn^{2+} + 2e^-$$

This occurs because zinc is a more reactive metal than iron and therefore zinc forms positive ions more readily than iron does.

The electrons travel from the zinc to the iron. The iron does not lose electrons, which means that oxidation of iron (which is the first stage of rusting) does not occur.

If tin is used instead of zinc, when the tin is scratched the iron will be oxidised in preference to the tin because iron is a more reactive metal than tin. Thus tin only prevents rusting when it is not scratched.

Common error

- It is a common error for students to state that if galvanised iron or steel is scratched and exposed to air and water, the zinc forms rust instead of iron. Iron is the only metal that can form rust.

● Nitrogen and fertilisers

Fertilisers are substances that are added to the soil to supply nutrients that are essential for the growth of plants.

Fertilisers contain nitrogen, phosphorus and potassium (NPK).

Very few plants can utilise nitrogen from the air. The nitrogen must be supplied to the plants in the form of fertilisers containing ammonium salts or nitrates.

The Haber process

Nitrogen obtained from the fractional distillation of liquid air (see earlier in this Chapter) and hydrogen from hydrocarbons by cracking (see Chapter 14) or from steam, are first converted into ammonia in the Haber process.

Nitrogen and hydrogen react together in a reversible reaction to produce ammonia.

$$N_{2(g)} + 3H_{2(g)} \rightleftharpoons 2NH_{3(g)}$$

The gases are

- passed over a catalyst of iron
- at a temperature of $450\,°C$
- at a pressure of 200 atmospheres.

The mixture emerging from the catalyst chamber contains about 15% ammonia. The ammonia is separated from the unreacted nitrogen and hydrogen by liquefying it, and the unreacted nitrogen and hydrogen are passed over the catalyst again. Eventually all the nitrogen and hydrogen are converted into ammonia.

The ammonia is used for

- the manufacture of fertilisers, such as ammonium sulfate and ammonium phosphate
- the manufacture of nitric acid which is converted into fertilisers, such as ammonium nitrate.

Exam-style questions

1 The percentage of oxygen in the air can be determined by passing air backwards and forwards over heated copper, using the apparatus shown in Figure 11.3. The copper was in excess.

The volume of air at the start was $100\,cm^3$. As the air was passed backwards and forwards, the volume of air decreased. The final volume, measured at room temperature and pressure, was $79\,cm^3$.

  a What colour change involving the copper would you expect to see? [2 marks]

  b Write a chemical equation for the reaction in the heated tube. [2 marks]

  c Why does the volume of air decrease? [1 mark]

d What is meant by the phrase 'the copper was in excess'? [1 mark]

e Identify the main gas present in the 79 cm³ remaining
  at the end. [1 mark]
  [Total: 7 marks]

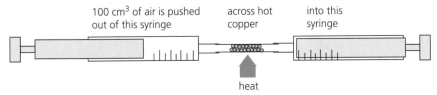

100 cm³ of air is pushed out of this syringe | across hot copper | into this syringe

heat

Figure 11.3

2 The equation for the equilibrium occurring in the Haber process is

$$N_{2(g)} + 3H_{2(g)} \rightleftharpoons 2NH_{3(g)}$$

The forward reaction is exothermic.

The reaction is carried out by passing nitrogen and hydrogen over a catalyst
of iron at a temperature of 450 °C and a pressure of 200 atmospheres.

a What is meant by the term catalyst? [2 marks]

b Copy and complete the table.

	Yield of ammonia		Rate of production of ammonia	
Higher temperature than 450 °C		[1 mark]		[1 mark]
Higher pressure than 200 atmospheres		[1 mark]		[1 mark]
Not using a catalyst		[1 mark]		[1 mark]

c Ammonia is converted into fertilisers such as ammonium sulfate.

  i Which substance would be added to ammonia to produce
    ammonium sulfate? [1 mark]

  ii Write a chemical equation for the reaction occurring in
     question (c)(i) above. [2 marks]

d Ammonia is also converted into the fertiliser ammonium phosphate.
  Complete the equation for the formation of ammonium phosphate
  from ammonia. (You may leave out the state symbols). [2 marks]
  [Total: 13 marks]

  $$NH_3 + H_3PO_4 \rightarrow$$

Sulfur

Key objectives

By the end of this section, you should be able to

- name some sources of sulfur
- know that sulfur is used in the manufacture of sulfuric acid
- know that sulfur dioxide is used to bleach wood pulp in the manufacture of paper, and as a food preservative (by killing bacteria)

- describe the manufacture of sulfuric acid by the Contact process, including essential conditions and reactions
- describe the properties and uses of dilute and concentrated sulfuric acid.

● Key terms

Sulfuric acid	Used for making detergents and fertilisers, used as the acid in car batteries and used for making paints, dyes and synthetic fibres
Dilute sulfuric acid	A typical strong acid
Sulfur dioxide	Used for bleaching of wood pulp in the manufacture of paper and in the preservation of food by killing bacteria

Sulfur is found as the impure element in sulfur beds below the ground. It is also found as metallic ores, mainly sulfides, such as zinc blende (see Chapter 10).

● Manufacture of sulfuric acid

Sulfur is used in the manufacture of **sulfuric acid** in the Contact process. The sulfur is first converted into **sulfur dioxide**, $SO_{2(g)}$, by heating it strongly in air.

$$S_{(s)} + O_{2(g)} \rightarrow SO_{2(g)}$$

The sulfur dioxide is then mixed with oxygen and the two gases are

- passed over a catalyst of vanadium(v) oxide, $V_2O_{5(s)}$
- at a temperature of $450\,°C$
- at a pressure of 1–2 atmospheres

producing an equilibrium mixture containing the two reactant gases and sulfur trioxide, $SO_{3(g)}$.

$$2SO_{2(g)} + O_{2(g)} \rightleftharpoons 2SO_{3(g)}$$

Sulfur trioxide is then dissolved in 98% concentrated sulfuric acid to produce oleum, $H_2S_2O_{7(l)}$.

$$SO_{3(g)} + H_2SO_{4(l)} \rightarrow H_2S_2O_{7(l)}$$

The oleum is then added to the correct amount of water to produce sulfuric acid of the required concentration.

$$H_2S_2O_{7(l)} + H_2O_{(l)} \rightarrow 2H_2SO_{4(l)}$$

Examiner's tip

If sulfur trioxide is added to water, a reaction occurs and sulfuric acid is the product.

$SO_{3(g)} + H_2O_{(l)} \rightarrow H_2SO_{4(l)}$

However, this reaction is very exothermic and the heat given off is sufficient to vaporise the sulfuric acid and a thick mist is produced that is very difficult to collect. Thus the process is not carried out in this way.

Cambridge IGCSE Chemistry Study and Revision Guide © David Besser

● Uses of sulfuric acid

Sulfuric acid is used

- for making detergents
- for making fertilisers (see Chapter 11)
- as the acid in car batteries
- for making paints, dyes and synthetic fibres.

● Properties of dilute sulfuric acid

Dilute sulfuric acid is a typical strong acid.

It is a diprotic acid, which means that one molecule of sulfuric acid releases two protons when it forms ions in aqueous solution.

$$H_2SO_{4(aq)} \rightarrow 2H^+_{(aq)} + SO_4^{2-}_{(aq)}$$

It can also release one proton.

$$H_2SO_{4(aq)} \rightarrow H^+_{(aq)} + HSO_4^-_{(aq)}$$

Therefore, sulfuric acid can form salts called **sulfates** containing SO_4^{2-} and acid salts called **hydrogen sulfates**, containing HSO_4^- (see Chapter 8).

It reacts

- with metals above hydrogen in the reactivity series to produce a salt and hydrogen (see Chapters 8 and 10)
- with bases (including alkalis) to form salts (sulfates) and acid salts (hydrogen sulfates), and water (see Chapter 8)
- with carbonates to form a salt, water and carbon dioxide (see Chapter 8).

● Uses of sulfur dioxide

Sulfur dioxide is used

- for the bleaching of wood pulp in the manufacture of paper
- in the preservation of food by killing bacteria.

Exam-style questions

1 Sulfuric acid is made in the Contact process. The catalysed reaction

$$2SO_{2(g)} + O_{2(g)} \rightleftharpoons 2SO_{3(g)}$$

takes place at a temperature of 450 °C and a pressure of 1–2 atmospheres.

The forward reaction is exothermic.

a Name the catalyst in the above reaction.   [1 mark]

b Suggest why a temperature below 450 °C is not used.   [1 mark]

c What would happen to the rate at which equilibrium would be reached if a higher pressure was used? Explain your answer.   [2 marks]

Cambridge IGCSE Chemistry Study and Revision Guide © David Besser

d Suggest two reasons why a pressure above 1–2 atmospheres is not used. [2 marks]

e How is the sulfur trioxide produced in the above reaction converted into sulfuric acid? Give equations for any reactions that you refer to. [2 marks]

[Total: 8 marks]

2 Write chemical equations for the reactions between dilute sulfuric acid and

a copper(ii) carbonate [2 marks]

b aqueous sodium hydroxide in which sodium sulfate is produced [2 marks]

c aqueous potassium hydroxide in which potassium hydrogen sulfate is produced [1 mark]

d zinc. [1 mark]

[Total: 6 marks]

Cambridge IGCSE Chemistry Study and Revision Guide © David Besser

(13) Inorganic carbon chemistry

Key objectives

By the end of this section, you should be able to

- state that carbon dioxide and methane are greenhouse gases and explain how they may contribute to climate change
- state the formation of carbon dioxide in a variety of reactions
- state the sources of methane, including decomposition of vegetation and waste gases from digestion in animals

- describe the manufacture of quicklime (calcium oxide) from calcium carbonate (limestone) in terms of thermal decomposition
- name some uses of quicklime and slaked lime
- name the uses of calcium carbonate
- describe the carbon cycle in outline to include the processes of combustion, respiration and photosynthesis.

● Key terms

Greenhouse gases	Gases which absorb infrared radiation produced by the solar warming of the Earth's surface
Greenhouse effect	Greenhouse gases preventing heat energy from escaping from the Earth

● Carbon dioxide and methane

Carbon dioxide and methane are both **greenhouse gases**. This means that their bonds absorb infrared radiation produced by the solar warming of the Earth's surface. This prevents heat energy from escaping from the Earth. This effect, known as the **greenhouse effect**, is thought to lead to global warming.

Carbon dioxide is formed in laboratory reactions and in industrial and environmental processes

- by complete combustion of carbon-containing substances such as fossil fuels (see Chapter 14)
- as a product of the reaction between an acid and a carbonate (see Chapter 8)
- from the thermal decomposition of carbonates (see Chapter 10)
- by respiration in which glucose is oxidised to carbon dioxide and water in living things.

$$C_6H_{12}O_{6(aq)} + 6O_{2(g)} \rightarrow 6CO_{2(g)} + 6H_2O_{(l)}$$

Carbon dioxide is removed from the atmosphere by reaction with water in the leaves of green plants containing chlorophyll, in the presence of sunlight. The reaction is known as **photosynthesis**. The products are glucose and oxygen.

$$6CO_{2(g)} + 6H_2O_{(l)} \rightarrow C_6H_{12}O_{6(aq)} + 6O_{2(g)}$$

Methane is formed by

- decomposition of vegetation
- Waste gases from digestion in animals.

Examiner's tip
The equations for respiration and photosynthesis are the reverse of one another.

The carbon cycle shows how the percentage of carbon dioxide in the atmosphere remains approximately constant at 0.03% due to the various processes by which it is released into and absorbed from the atmosphere.

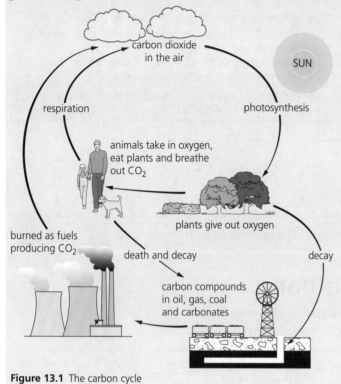

Figure 13.1 The carbon cycle

● Carbonates

Quicklime (calcium oxide) is manufactured by the thermal decomposition of calcium carbonate (limestone). This is a reversible reaction.

$$CaCO_{3(s)} \rightleftharpoons CaO_{(s)} + CO_{2(g)}$$

The process is carried out in a lime kiln. A draft of air carries out the carbon dioxide and causes the equilibrium to shift to the right (see Chapter 7). Eventually all the calcium carbonate is converted into calcium oxide.

Calcium oxide reacts with water to produce calcium hydroxide (slaked lime).

$$CaO_{(s)} + H_2O_{(l)} \rightarrow Ca(OH)_{2(s)}$$

Uses of quick lime and slaked lime

Both quicklime, CaO, and slaked lime, $Ca(OH)_2$, can be used to

● neutralise acidic soil (see Chapter 8)
● neutralise acidic industrial waste products, e.g. flue gas desulfurisation, in which sulfur dioxide gas in factory emissions is neutralised by quicklime or slaked lime.

Uses of limestone

Limestone, $CaCO_3$, is used in the

● manufacture of iron (see Chapter 10)
● manufacture of cement.

Exam-style questions

1 State why it is necessary to neutralise soil acidity and give the
chemical name of a compound which can be added to soil in
order to neutralise it.   [Total: 2 marks]

2 a Carbon dioxide can be produced by the

   i complete combustion of octane

   ii thermal decomposition of calcium carbonate

   iii reduction of iron(III) oxide by carbon monoxide

   iv reaction between calcium carbonate and dilute hydrochloric acid.

   Write a chemical equation for each of these reactions.
   You may omit state symbols.   [Total: 7 marks]

   b Carbon dioxide can be removed from the Earth's atmosphere
   by photosynthesis.

   i Name the other reactant in photosynthesis.   [1 mark]

   ii Name the two products of photosynthesis.   [2 marks]

   iii State two conditions that are required for
   photosynthesis.   [2 marks]
                                                          [Total: 12 marks]

Cambridge IGCSE Chemistry Study and Revision Guide © David Besser

Organic chemistry 1

● Key terms

Empirical formula	The smallest whole number ratio of the atoms of each element in a compound
Molecular formula	This is the number of atoms of each element in one molecule of a substance
Displayed formula	All the atoms and all the bonds in one molecule of the compound
Structural formula	This shows how atoms are arranged in groups of atoms
Structural isomerism	The existence of compounds with the same molecular formula but different structural formulae

Organic compounds are covalent compounds containing carbon atoms bonded to hydrogen (always), as well as oxygen, the halogens and nitrogen atoms.

● Homologous series

Organic compounds belong to 'families' of similar compounds known as **homologous series**, examples of which are alkanes, alkenes, alcohols, carboxylic acids and esters.

Homologous series are compounds which have

- the same general formula: this is because each member differs from the previous member by a $-CH_2-$ group of atoms
- the same chemical properties: this is because each member has the same functional group
- physical properties that show a constant gradation, e.g. melting points and boiling points that show almost constant increases.

A **functional group** is a group of atoms that all members of a homologous series have in common and which is responsible for all members of a homologous series having the same chemical properties.

Formulae of organic compounds

Organic compounds have several different types of formulae (Table 14.1). These are

- **Empirical formula.** This is the smallest whole number ratio of the atoms of each element in a compound (see Chapter 4).
- **Molecular formula.** This is the number of atoms of each element in one molecule of a substance (see Chapter 4). This gives no information about how the atoms are joined together.
- **Displayed formula.** This shows all the atoms and all the bonds in one molecule of the compound.
- **Structural formula.** This shows how atoms are arranged in groups of atoms.

Examiner's tip

The term 'displayed formula' is not always used in examination questions. Instead students are more likely to be asked to draw the structure of a molecule showing all atoms and all bonds.

Table 14.1 Formulae of organic compounds using butane as an example

Compound	Empirical formula	Molecular formula	Displayed formula	Structural formula
Butane	C_2H_5	C_4H_{10}		$CH_3CH_2CH_2CH_3$

Structural isomerism

Structural isomerism is the existence of compounds with the same molecular formula but different structural formulae (and therefore different displayed formulae).

Example

There are two structural isomers with the molecular formula C_4H_{10}. These have different structural and displayed formulae as shown in Table 14.2. Because they are two different compounds they have two different names.

Butane is often referred to as a straight chain or an unbranched molecule, because the carbon atoms are arranged one after another.

2-methylpropane is often referred to as a branched chain molecule.

Table 14.2 Structural isomerism

Molecular formula	C_4H_{10}	C_4H_{10}
Displayed formula		
Structural formula	$CH_3CH_2CH_2CH_3$	CH_3CHCH_3 $\|$ CH_3 or $CH_3CH(CH_3)CH_3$
Name	Butane	2-methylpropane

Common errors

Students often confuse the two words isotopes and isomers:

- **Isotopes** (see Chapter 3) are atoms of the same element with the same proton number but different nucleon number.
- Structural **isomers** are compounds with the same molecular formula but different structural formulae.

Allotrope is another similar word which you may have come across. This refers to different crystalline forms of the same element, e.g. diamond and graphite.

- When drawing displayed formulae, all carbon atoms must have four bonds (sticks) and all hydrogen atoms must have one bond. Oxygen atoms must have two bonds, halogen atoms one bond and nitrogen atoms three bonds.

Hydrocarbons

Hydrocarbons are compounds containing carbon and hydrogen *only*. Alkanes and alkenes are important examples of hydrocarbons.

Alkanes

- **Alkanes** are members of a homologous series with the general formula C_nH_{2n+2}.
- The names of alkanes all end in **-ane**.
- Alkanes are *saturated* hydrocarbons; this means that all their bonds are single bonds (either C–C or C–H).
- Alkanes do not contain a functional group, because they contain only C–C and C–H bonds which can be said of all other homologous series of organic compounds.

Table 14.3 First four unbranched members of the homologous series of alkanes

No.	Name	Molecular formula	Structural formula
1	Methane	CH_4	CH_4
2	Ethane	C_2H_6	CH_3CH_3
3	Propane	C_3H_8	$CH_3CH_2CH_3$
4	Butane	C_4H_{10}	$CH_3CH_2CH_2CH_3$

> **Examiner's tip**
>
> The names of alkanes are important because unbranched members of all other homologous series are named after the alkane with the same number of carbon atoms. Therefore all organic molecules with
>
> - **1** carbon atom begin with **meth-**
> - **2** carbon atoms begin with **eth-**
> - **3** carbon atoms begin with **prop-**
> - **4** carbon atoms begin with **but-**.
>
> Some examples of organic compounds with two carbon atoms are
>
> | Alkane | Alkene | Alcohol | Carboxylic acid | Chloroalkane |
> |---|---|---|---|---|
> | **Eth**ane | **Eth**ene | **Eth**anol | **Eth**anoic acid | Chloro**eth**ane |

Properties of alkanes

Alkanes are unreactive compared to alkenes. This is because the single bonds in alkanes need a lot of energy to break, but the double bonds in alkenes need less energy to be converted into single bonds (which is what happens to alkenes in **addition reactions**).

Examiner's tip

When asked 'What is meant by the term hydrocarbon?' it is important to use the word 'only', as well as referring to compounds containing carbon and hydrogen. Compounds such as ethanol, C_2H_5OH, contain carbon and hydrogen, but because they also contain oxygen, they are not hydrocarbons.

Cambridge IGCSE Chemistry Study and Revision Guide © David Besser

Combustion

Alkanes undergo **combustion** in air or oxygen producing energy, which is why alkanes are used as fuels.

Complete combustion occurs in excess oxygen. The products are carbon dioxide and water, for example

$$CH_{4(g)} + 2O_{2(g)} \rightarrow CO_{2(g)} + 2H_2O_{(l)}$$

Incomplete combustion of alkanes in a limited supply of air or oxygen leads to the production of (toxic) carbon monoxide as well as water (see Chapter 11).

$$2CH_{4(g)} + 3O_{2(g)} \rightarrow 2CO_{(g)} + 4H_2O_{(l)}$$

Reaction with chlorine

It is not possible to add atoms to alkane molecules without first removing atoms. Therefore alkanes undergo **substitution reactions** as opposed to **addition reactions**.

A **substitution** reaction is a reaction in which one atom or group of atoms is replaced by another atom or group of atoms.

When methane is reacted with chlorine in the presence of ultraviolet light, a substitution reaction occurs in which one chlorine atom replaces one hydrogen atom in methane. The organic product is chloromethane.

$$CH_{4(g)} + Cl_{2(g)} \rightarrow CH_3Cl_{(l)} + HCl_{(g)}$$

Unless the chlorine supply is limited, the reaction should not be used as a method of preparation of chloromethane because the chloromethane reacts with more chlorine. All the hydrogen atoms are substituted by chlorine atoms, one at a time, until all the hydrogen atoms have been replaced by chlorine atoms. Hydrogen chloride gas is produced at each stage.

$$CH_3Cl_{(l)} + Cl_{2(g)} \rightarrow CH_2Cl_{2(l)} + HCl_{(g)}$$
$$\text{dichloromethane}$$

$$CH_2Cl_{2(l)} + Cl_{2(g)} \rightarrow CHCl_{3(l)} + HCl_{(g)}$$
$$\text{trichloromethane}$$

$$CHCl_{3(l)} + Cl_{2(g)} \rightarrow CCl_{4(l)} + HCl_{(g)}$$
$$\text{tetrachloromethane}$$

Similar reactions occur with other alkanes and chlorine.

● Alkenes

- **Alkenes** are members of a homologous series with the general formula C_nH_{2n}.
- Alkenes contain a functional group which is a $C = C$.
- Because a $C = C$ must be present in all alkenes, there is no alkene with one carbon atom.
- The names of alkenes all end in **–ene**.
- Alkenes are *unsaturated* hydrocarbons.
- Not all the bonds are single bonds in unsaturated hydrocarbons. Unsaturated molecules contain at least one carbon–carbon double bond or carbon–carbon triple bond.

Cambridge IGCSE Chemistry Study and Revision Guide © David Besser

Testing for unsaturation

Aqueous bromine (bromine water) can be used to distinguish between saturated and unsaturated substances (Table 14.4).

Table 14.4 Using bromine water to test for saturated and unsaturated substances

Type of molecule	Saturated	Unsaturated
Effect of adding aqueous bromine	No change (aqueous bromine remains pale brown)	Aqueous bromine changes from pale brown to colourless

Structural isomerism in alkenes

There is only one structure for the alkenes containing two and three carbon atoms: ethene and propene.

With four carbon atoms (C_4H_8) there are two unbranched alkenes, because the double bond can be in two different positions in the carbon chain (Table 14.5).

Table 14.5

Molecular formula	C_4H_8	C_4H_8
Displayed formula	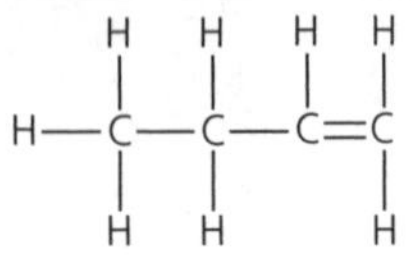	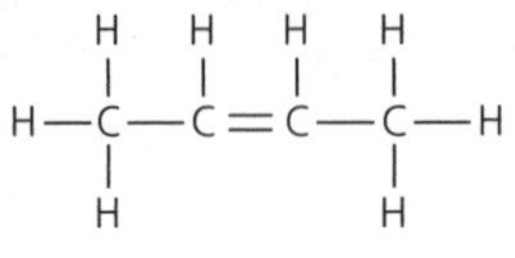
Structural formula	$CH_3CH_2CH = CH_2$	$CH_3CH = CHCH_3$
Name	But-1-ene	But-2-ene

The number **1** in but-**1**-ene means that the double bond is between carbon atoms **1** and 2.

The number **2** in but-**2**-ene means that the double bond is between carbon atoms **2** and 3.

Structures of unbranched alkenes containing up to four carbon atoms are shown in Table 14.6.

Table 14.6

No.	Molecular formula	Name	Structural formula
2	C_2H_4	Ethene	$CH_2 = CH_2$
3	C_3H_6	Propene	$CH_3CH = CH_2$
4	C_4H_8	But-1-ene	$CH_3CH_2CH = CH_2$
4	C_4H_8	But-2-ene	$CH_3CH = CHCH_3$

Examiner's tip

When students are asked to give the structures of two alkenes with molecular formula C_4H_8, they often draw

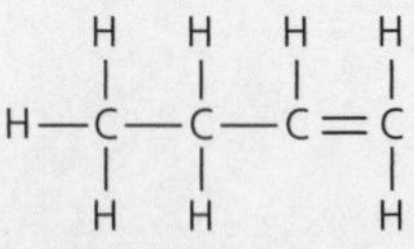

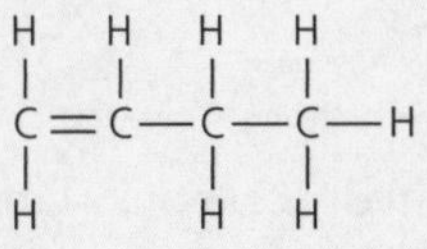

Figure 14.1

These are both but-1-ene drawn differently. One is the same as the other, only it is drawn back to front. The double bond is between carbon atoms 1 and 2 in both cases.

Isomers must be different molecules, not the same molecule drawn differently.

Cambridge IGCSE Chemistry Study and Revision Guide © David Besser

Reactions of alkenes

It is possible to add atoms to alkene molecules without first removing atoms.
Therefore alkanes undergo **addition** reactions as opposed to substitution reactions.

Addition means two molecules join together to make one molecule.

In the addition reactions of alkenes, the double bond becomes a single
bond and an atom or a group of atoms joins on to both carbon atoms that
formed the double bond.

$$\diagup_{\diagdown} C = C \diagup_{\diagdown} \quad X - Y \quad \longrightarrow \quad X - \diagup_{\diagdown} C - C \diagup_{\diagdown} - Y$$

Figure 14.2 An addition reaction

With bromine

If bromine the element ($Br_{2(l)}$) or aqueous bromine ($Br_{2(aq)}$) is added to any
alkene, an addition reaction occurs. If the alkene is ethene, the product is
1, 2-dibromoethane.

$$CH_2 = CH_2 + Br_2 \rightarrow CH_2BrCH_2Br$$

With hydrogen (industrial)

If ethene and hydrogen are passed over a nickel catalyst at 200 °C, the
product is ethane. This addition reaction is called **hydrogenation**.

$$CH_2 = CH_2 + H_2 \rightarrow CH_3CH_3$$

Hydrogenation is used in industry to convert (unsaturated) vegetable oils
to (saturated) margarine by reaction with hydrogen in the presence of a nickel
catalyst.

With steam (industrial)

Ethene reacts with steam using a catalyst of phosphoric acid (H_3PO_4), at
300 °C and 60 atmospheres pressure. The product is ethanol.

$$CH_2 = CH_2 + H_2O \rightleftharpoons CH_3CH_2OH$$

This reaction, known as hydration of ethene, is used to manufacture
ethanol (see Chapter 15).

The reactions of ethene are summarised in Figure 14.3.

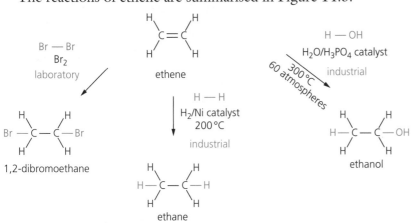

Figure 14.3 The reactions of ethene

Cambridge IGCSE Chemistry Study and Revision Guide © David Besser

Fuels and the petroleum industry

Fossil fuels are fuels formed by natural processes over millions of years as a result of the decay of buried dead organisms. Examples are **coal**, **natural gas** and **petroleum** (crude oil). Fossil fuels are known as non-renewable fuels because once they run out they cannot be replaced. They are also known as a finite resource.

Methane, CH_4, is the main constituent of natural gas.

Petroleum (crude oil) is a mixture of hydrocarbons which is separated by fractional distillation. The process does not produce individual hydrocarbons, but instead produces mixtures of hydrocarbons known as **fractions**. These fractions (Figure 14.4) are mixtures of hydrocarbons which have a narrow range of boiling points. As the boiling point ranges increase, the hydrocarbons contain an increasing number of carbon atoms.

Manufacture of alkenes by cracking

Fractional distillation of petroleum produces

- higher boiling point fractions that are in excess of requirement
- insufficient amounts of the lower boiling point fractions, which are particularly in demand as
  - fuels for petrol engines (alkanes between C_5 and C_{10})
  - monomers (short-chain alkenes) for the production of polymers
  - sources of organic chemicals.

Alkenes are manufactured by **cracking** long-chain alkanes obtained from petroleum. Cracking refers to decomposition of alkanes which specifically involves breaking carbon–carbon bonds to form smaller molecules. There are two types of cracking: catalytic and thermal.

Catalytic cracking

The alkane molecules are passed over catalysts known as zeolites at temperatures of $500\,°C$.

Thermal cracking

This uses a higher temperature ($800\,°C$) than catalytic cracking.

Cracking produces
- short-chain alkenes for production of polymers and organic chemicals
- alkanes containing between 5 and 10 carbon atoms as fuels for petrol engines
- hydrogen which is used to manufacture ammonia.

If a long-chain alkane is cracked, different molecules of the alkane may break in different places to give a variety of products which can also be separated by fractional distillation.

For example, $C_{14}H_{30}$ molecules could crack into octane and propene

$$C_{14}H_{30} \rightarrow C_8H_{18} + 2C_3H_6$$

or ethene, propene and hydrogen.

$$C_{14}H_{30} \rightarrow C_2H_4 + 4C_3H_6 + H_2$$

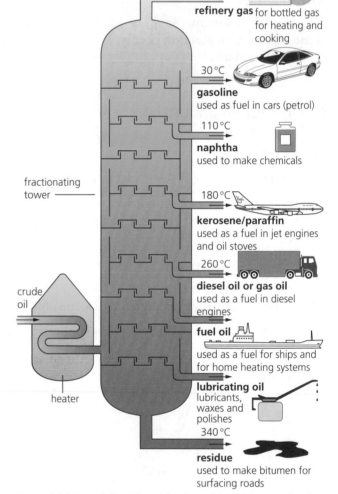

refinery gas for bottled gas for heating and cooking

$30\,°C$

gasoline used as fuel in cars (petrol)

$110\,°C$

naphtha used to make chemicals

fractionating tower

$180\,°C$

kerosene/paraffin used as a fuel in jet engines and oil stoves

$260\,°C$

diesel oil or gas oil used as a fuel in diesel engines

crude oil

fuel oil used as a fuel for ships and for home heating systems

lubricating oil lubricants, waxes and polishes

$340\,°C$

heater

residue used to make bitumen for surfacing roads

Figure 14.4 Uses of the different fractions obtained from crude oil

Examiner's tip

Students will not be asked to predict what the definite products of cracking are without being provided with further information.

● Sample exam-style question

Write an equation for the cracking of hexane into an alkane and an alkene, both having the same number of carbon atoms.

Answer

Both

$$C_6H_{14} \rightarrow C_3H_8 + C_3H_6$$

and

$$C_6H_{14} \rightarrow C_2H_6 + 2C_2H_4$$

are acceptable answers, because both produce an alkane and an alkene with the same number of carbon atoms. Neither answer is more correct than the other.

If the question had specified a 1:1 mole ratio of the products, only $C_6H_{14} \rightarrow C_3H_8 + C_3H_6$ would be correct.

Exam-style questions

1   Draw the structures, showing all the atoms and all the bonds, of two different unbranched alkenes with the molecular formula C_5H_{10}. You are not expected to name the alkenes.   [2 marks]

2   Ethane reacts with chlorine in a substitution reaction.

   a   Under what condition does the reaction take place?   [1 mark]

   b   Name the organic product formed when ethane and chlorine react in a 1:1 mole ratio.   [1 mark]

   c   If an excess of chlorine is used, give the molecular formula of one other organic product that could form.   [1 mark]
                                                    [Total: 3 marks]

3   Propene reacts with

    i   aqueous bromine   **ii**   hydrogen   **iii**   steam.

   a   What name is given to the type of reaction that occurs in all three cases?   [1 mark]

   b   State the observation you would expect to see in reaction (i) if excess propene is used.   [2 marks]

   c   Write down the molecular formulae of the products that form in reactions (i), (ii) and (iii).   [3 marks]

   d   What is the name of the catalyst used in (ii)?   [1 mark]
                                                    [Total: 7 marks]

4   Alkanes are converted into alkenes by cracking.

   a   What is the molecular formula of the alkane that contains nine carbon atoms?   [1 mark]

   b   Draw the structure and give the name of an alkene with four carbon atoms.   [2 marks]

   c   Write an equation for the cracking of octane, C_8H_{18}, into

    i   an alkane and an alkene formed in the mole ratio 1:2

    ii   hydrogen and two other products.   [2 marks]
                                                     [Total: 5 marks]

Cambridge IGCSE Chemistry Study and Revision Guide © David Besser

Key objectives

By the end of this section, you should be able to

- describe the manufacture of ethanol by fermentation and by the catalytic addition of steam to ethene
- outline the advantages and disadvantages of these two methods of manufacturing ethanol

- describe the properties of ethanol in terms of burning
- name the uses of ethanol as a solvent and as a fuel
- name and draw the structures of unbranched alcohols and carboxylic acids containing up to four carbon atoms per molecule
- name and draw the structures of the esters which can be made from unbranched alcohols and carboxylic acids, each containing up to four carbon atoms

- describe the properties of aqueous ethanoic acid

- describe the formation of ethanoic acid by the oxidation of ethanol with acidified potassium manganate(VII)
- describe ethanoic acid as a typical weak acid
- describe the reaction of a carboxylic acid with an alcohol in the presence of a catalyst to give an ester

- define polymers as large molecules built up from small molecules called monomers
- explain the differences between addition and condensation polymerisation and understand that different polymers have different units and/or different linkages

- describe the formation of polyethene as an example of addition polymerisation
- name some uses of plastics and of man-made fibres such as nylon and Terylene
- describe the pollution problems caused by non-biodegradable plastics

- deduce the structure of the addition polymer from a given alkene and vice versa
- describe the formation of nylon (a polyamide) and Terylene (a polyester) by condensation polymerisation

- name proteins and carbohydrates as constituents of food
- describe the structure of proteins as possessing the same amide linkages as nylon, but with different units
- describe the hydrolysis of proteins to amino acids
- describe complex carbohydrates in terms of a large number of sugar units joined together by condensation polymerisation
- describe hydrolysis of complex carbohydrates (e.g. starch) by acids or enzymes to give simple sugars
- describe in outline the usefulness of chromatography in separating and identifying the products of hydrolysis of carbohydrates and of proteins.

● Key terms

Ethanol	An alcohol used as a fuel (e.g. in spirit camping stoves and in petrol)
Polymers	Large molecules made up by the reactions of small molecules called monomers
Polymerisation	The formation of polymers from monomers

● Alcohols

Manufacture of ethanol

Ethanol is manufactured on a large scale by fermentation of carbohydrates and hydration of ethene.

Fermentation of carbohydrates

Carbohydrates, such as sugar, are broken down by enzymes in yeast to produce glucose, $C_6H_{12}O_6$. The enzymes in yeast also catalyse the breakdown of glucose to form ethanol and carbon dioxide. The reaction occurs at a temperature of $37\,°C$ and is carried out under anaerobic conditions (in the absence of oxygen).

$$C_6H_{12}O_{6(aq)} \rightarrow 2C_2H_5OH_{(aq)} + 2CO_{2(g)}$$

When the concentration of ethanol reaches 14%, it kills the yeast. The yeast cells are removed by filtration and the ethanol is purified by fractional distillation.

Hydration of ethene

Ethene is produced from petroleum by fractional distillation followed by cracking of long-chain alkanes.

Ethene reacts with steam using a catalyst of phosphoric acid, H_3PO_4, at 300 °C and 60 atmospheres pressure.

$$CH_{2(g)}=CH_2 + H_2O_{(g)} \rightleftharpoons C_2H_5OH_{(g)}$$

The advantages and disadvantages of the two processes are shown in Table 15.1.

Table 15.1 Advantages and disadvantages of fermentation and hydration in the manufacture of ethanol

	Fermentation	Hydration
Advantages	Uses carbohydrates from plants which are a renewable resource	There is only one product in the reaction which means there is no waste
	Requires a temperature of 37 °C which means energy costs are low	A continuous flow process is used which is efficient
Disadvantages	A batch process is used which is inefficient	Uses ethene from petroleum which is a non-renewable resource
	Land which could be used to grow plants for food is used for ethanol production	Requires a temperature of 300 °C which means energy costs are high

Properties of ethanol

Ethanol is used as a fuel in spirit camping stoves and it is also added to petrol and used in ethanol fuel cells.

It undergoes complete combustion to produce carbon dioxide and water.

$$C_2H_5OH_{(l)} + 3O_{2(g)} \rightarrow 2CO_{2(g)} + 3H_2O_{(l)}$$

Ethanol is also used on a large scale as a solvent.

Conversion of alcohols to carboxylic acids

Carboxylic acids are formed in the laboratory by oxidation of alcohols using aqueous potassium manganate(VII) which acts as an oxidising agent when acidified with sulfuric acid.

Ethanol is oxidised to ethanoic acid by this method.

A simplified version of the equation, which represents oxygen from the oxidising agent as [O], is

$$CH_3CH_2OH + 2[O] \rightarrow CH_3COOH + H_2O$$

Formulae and names of alcohols

- Alcohols are members of a homologous series with the general formula $C_nH_{2n+1}OH$.
- The names of alcohols all end in -**ol**.
- Alcohols contain the functional group which is the −O−H group.
- Alcohols with more than two carbon atoms have unbranched structural isomers in which the O−H group can be in different positions on the carbon chain. A number is used to indicate the position of the O−H group (Table 15.2).

Cambridge IGCSE Chemistry Study and Revision Guide © David Besser

Unbranched alcohols with up to four carbon atoms are shown in Table 15.2.

Table 15.2

Number of carbon atoms	Displayed formula	Structural formula	Name
1	H | H—C—O—H | H	CH_3OH	Methanol
2	H H | | H—C—C—O—H | | H H	CH_3CH_2OH or C_2H_5OH	Ethanol
3	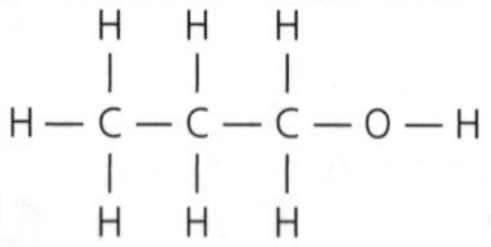 H H H | | | H—C—C—C—O—H | | | H H H	$CH_3CH_2CH_2OH$	Propan-1-ol
3	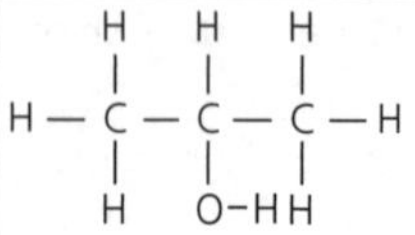 H H H | | | H—C—C—C—H | | | H O-H H	CH_3CHCH_3 | OH or $CH_3CH(OH)CH_3$	Propan-2-ol
4	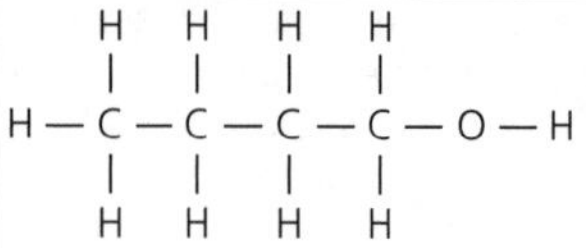 H H H H | | | | H—C—C—C—C—O—H | | | | H H H H	$CH_3CH_2CH_2CH_2OH$	Butan-1-ol
4	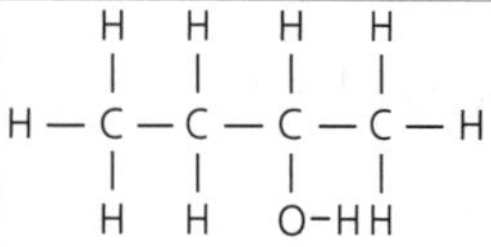 H H H H | | | | H—C—C—C—C—H | | | | H H O-H H	$CH_3CH_2CHCH_3$ | OH or $CH_3CH_2CH(OH)CH_3$	Butan-2-ol

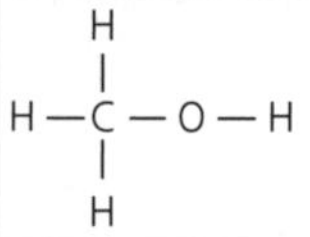

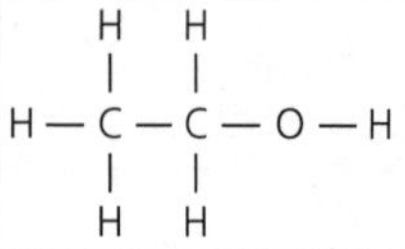

Carboxylic acids

- **Carboxylic acids** are members of a homologous series.
- The names of carboxylic acids all end in -**oic** acid.
- The functional group in carboxylic acids is –COOH which can also be written as $–CO_2H$.
- This is displayed as shown in Figure 15.1

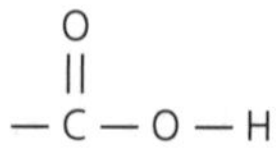

O
‖
—C—O—H

Figure 15.1

Unbranched carboxylic acids with up to four carbon atoms are shown in Table 15.3.

Table 15.3

Number of carbon atoms	Displayed formula	Structural formula	Name
1	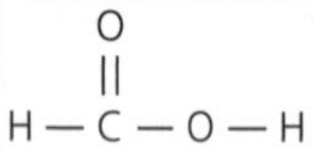 O ‖ H—C—O—H	HCOOH	Methanoic acid
2	H O | ‖ H—C—C—O—H | H	CH_3COOH	Ethanoic acid
3	H H O | | ‖ H—C—C—C—O—H | | H H	CH_3CH_2COOH	Propanoic acid
4	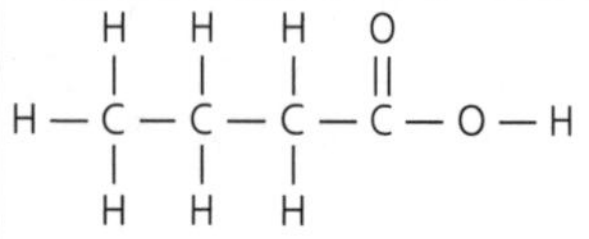 H H H O | | | ‖ H—C—C—C—C—O—H | | | H H H	$CH_3CH_2CH_2COOH$	Butanoic acid

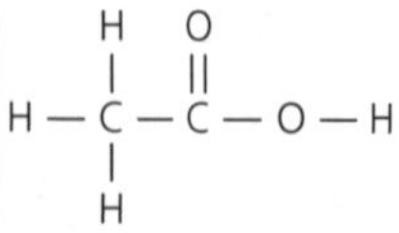

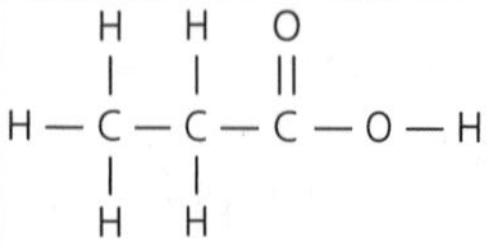

Cambridge IGCSE Chemistry Study and Revision Guide © David Besser

Properties of aqueous ethanoic acid

Ethanoic acid, CH_3COOH, is a typical weak acid. It reacts with metals, bases and carbonates to produce salts (see Chapter 8). The salts are called ethanoates and contain the ethanoate ion, CH_3COO^-.

● Esters

- **Esters** are sweet-smelling liquids.
- The names of esters all end in -**oate**.
- Esters have a general formula of $C_nH_{2n}O_2$.
- The functional group in esters is –COOR which can also be written as –CO$_2$R. R represents a group containing carbon and hydrogen atoms.
- The functional group in esters is displayed as shown in Figure 15.2.
- Esters are made by the reaction between a carboxylic acid and an alcohol.
- The type of reaction is called **esterification**. The alcohol and carboxylic acid are heated with a catalyst of concentrated sulfuric acid.

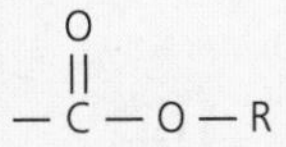

Figure 15.2

The general equation in words is:

carboxylic acid + alcohol $\rightleftharpoons$ ester + water

The molecules can be represented as

$$RCOOH + ROH \rightleftharpoons RCOOR + H_2O$$

The reaction occurs as

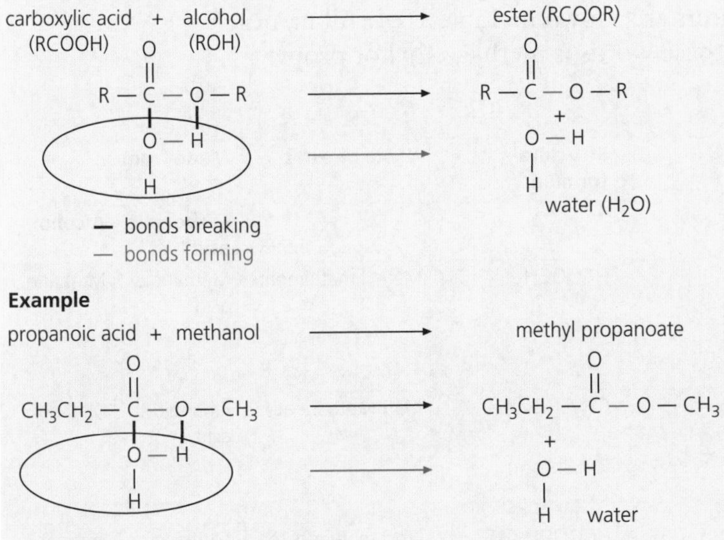

— bonds breaking
— bonds forming

Example

Figure 15.3 The formation of an ester

An example is

$$CH_3COOH_{(l)} + CH_3CH_2OH_{(l)} \rightleftharpoons CH_3COOCH_2CH_{3(l)} + H_2O_{(l)}$$
ethanoic acid + ethanol $\rightleftharpoons$ ethyl ethanoate + water

Naming esters is unlike naming any other organic molecules we have met so far. The formula is divided into two and each part is named according to the number of carbon atoms it contains as shown in Figure 15.4.

The names and formulae of unbranched esters are shown Table 15.4.

Cambridge IGCSE Chemistry Study and Revision Guide © David Besser

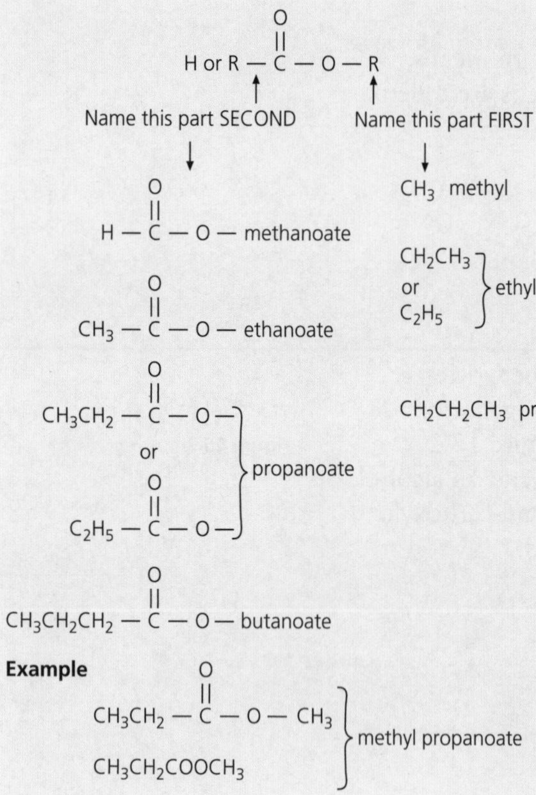

$$H \text{ or } R - \overset{\overset{\displaystyle O}{\|}}{C} - O - R$$

Name this part SECOND   Name this part FIRST

$$H - \overset{\overset{\displaystyle O}{\|}}{C} - O - \text{ methanoate}$$ CH_3 methyl

$$CH_3 - \overset{\overset{\displaystyle O}{\|}}{C} - O - \text{ ethanoate}$$ $\left.\begin{array}{l} CH_2CH_3 \\ \text{or} \\ C_2H_5 \end{array}\right\}$ ethyl

$$CH_3CH_2 - \overset{\overset{\displaystyle O}{\|}}{C} - O - \left.\begin{array}{l}\\ \text{or} \\ \end{array}\right.$$ $CH_2CH_2CH_3$ propyl
$$C_2H_5 - \overset{\overset{\displaystyle O}{\|}}{C} - O - \left.\begin{array}{l}\end{array}\right\} \text{propanoate}$$

$$CH_3CH_2CH_2 - \overset{\overset{\displaystyle O}{\|}}{C} - O - \text{ butanoate}$$

Example

$$\left.\begin{array}{l} CH_3CH_2 - \overset{\overset{\displaystyle O}{\|}}{C} - O - CH_3 \\[6pt] CH_3CH_2COOCH_3 \end{array}\right\} \text{methyl propanoate}$$

Figure 15.4 Naming esters

The names and formulae of unbranched esters are shown in Table 15.4. All names should be two combined words. The first of these words is methyl, ethyl or propyl.

Table 15.4 Names and formulae of unbranched esters

Number of carbon atoms	Molecular formula of ester	Displayed formula	Structural formula	Name of ester	Made from	
					Carboxylic acid	Alcohol
2	$C_2H_4O_2$	$H - \overset{\overset{O}{\|}}{C} - O - \overset{\overset{H}{\|}}{\underset{\underset{H}{\|}}{C}} - H$	$HCOOCH_3$	Methyl methanoate	Methanoic acid	Methanol
3	$C_3H_6O_2$	$H - \overset{\overset{H}{\|}}{\underset{\underset{H}{\|}}{C}} - \overset{\overset{O}{\|}}{C} - O - \overset{\overset{H}{\|}}{\underset{\underset{H}{\|}}{C}} - H$	CH_3COOCH_3	Methyl ethanoate	Ethanoic acid	Methanol
3	$C_3H_6O_2$	$H - \overset{\overset{O}{\|}}{C} - O - \overset{\overset{H}{\|}}{\underset{\underset{H}{\|}}{C}} - \overset{\overset{H}{\|}}{\underset{\underset{H}{\|}}{C}} - H$	$HCOOCH_2CH_3$	Ethyl methanoate	Methanoic acid	Ethanol
4	$C_4H_8O_2$	$H - \overset{\overset{H}{\|}}{\underset{\underset{H}{\|}}{C}} - \overset{\overset{H}{\|}}{\underset{\underset{H}{\|}}{C}} - \overset{\overset{O}{\|}}{C} - O - \overset{\overset{H}{\|}}{\underset{\underset{H}{\|}}{C}} - H$	$CH_3CH_2COOCH_3$	Methyl propanoate	Propanoic acid	Methanol
4	$C_4H_8O_2$	$H - \overset{\overset{H}{\|}}{\underset{\underset{H}{\|}}{C}} - \overset{\overset{O}{\|}}{C} - O - \overset{\overset{H}{\|}}{\underset{\underset{H}{\|}}{C}} - \overset{\overset{H}{\|}}{\underset{\underset{H}{\|}}{C}} - H$	$CH_3COOCH_2CH_3$	Ethyl ethanoate	Ethanoic acid	Ethanol
4	$C_4H_8O_2$	$H - \overset{\overset{O}{\|}}{C} - O - \overset{\overset{H}{\|}}{\underset{\underset{H}{\|}}{C}} - \overset{\overset{H}{\|}}{\underset{\underset{H}{\|}}{C}} - \overset{\overset{H}{\|}}{\underset{\underset{H}{\|}}{C}} - H$	$HCOOCH_2CH_2CH_3$	Propyl methanoate	Methanoic acid	Propan-1-ol

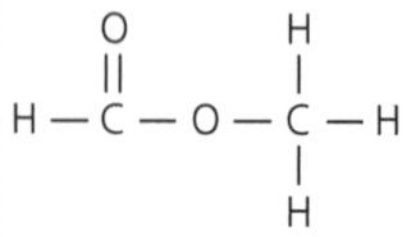

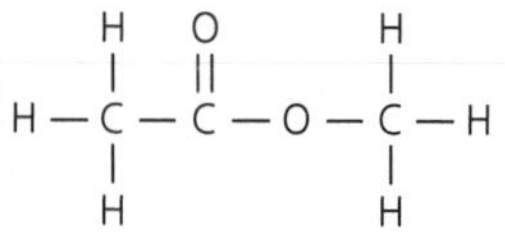

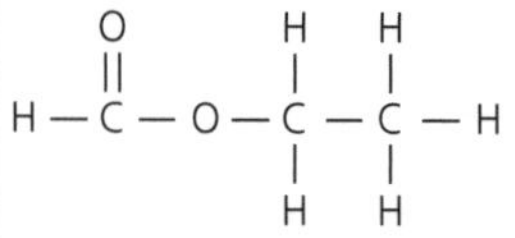

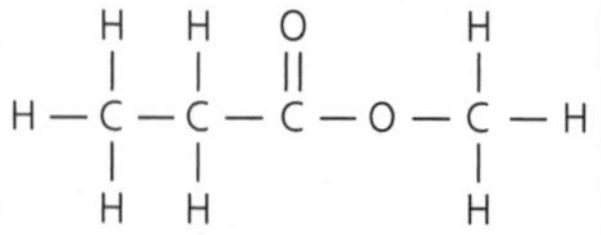

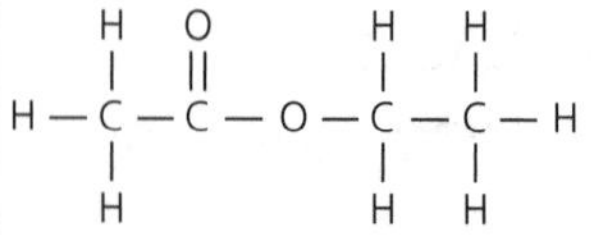

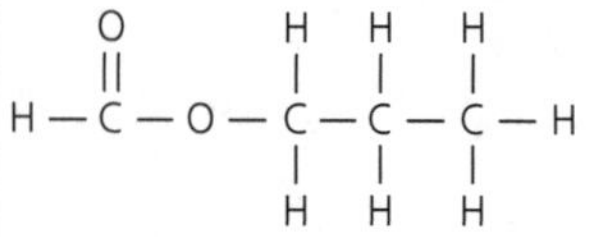

Cambridge IGCSE Chemistry Study and Revision Guide © David Besser

Examiner's tips

Remember that most formulae of organic compounds begin with a C atom. However, methanoic acid is written HCOOH and because it begins with an **H,** its structural formula is constantly written incorrectly by students. The same thing applies to methanoate esters, such as methyl methanoate which should be written **HCOOCH₃.**

- There are no esters with one carbon atom.
- There are two isomeric esters with three carbon atoms.
- There are three isomeric unbranched esters with four carbon atoms.

● Polymers

Polymers are large molecules (of no definite size) made up by the reactions of small molecules called **monomers**. The formation of polymers from monomers is called **polymerisation**.

There are two types of polymerisation known as **addition polymerisation** and **condensation polymerisation**.

Addition polymerisation

If alkenes such as ethene are treated to conditions of high temperature and high pressure in the presence of a suitable catalyst, the double bonds become single bonds, making more electrons available for the carbon atoms to join together and form long chains.

This happens to thousands of ethene molecules which join together to form one long-chain molecule. The chemical name of the product is polyethene. Its commercial name is polythene.

This reaction is known as addition polymerisation, because the monomers join together without the removal of any atoms. There is only one product just as in other addition reactions of alkenes (see Chapter 14).

The equation for polymerisation of ethene is shown in Figure 15.5 where n represents a number larger than 10 000.

ethene (monomer)   polyethene (polymer)

Figure 15.5

Other examples of addition polymerisation

Theoretically any molecule with a carbon–carbon double bond can form an addition polymer.

The chemical name of the polymer is always the same as that of the monomer with poly- in front.

Propene, $CH_3CH{=}CH_2$, undergoes addition polymerisation to form polypropene. The structure of the polymer can only be represented by drawing the monomer as in Figure 15.6, changing the double bond to a single bond and drawing two extension bonds on either side which show that the polymer extends in both directions.

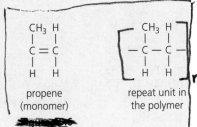

propene
(monomer)

repeat unit in
the polymer

Uses of addition polymers

Addition polymers have several uses which are dependent on strength, hardness, poor electrical and thermal conductivity. Examples are

- Polythene is used for carrier bags, buckets and bowls.
- PVC is used for window frames, guttering and insulating electrical wires.
- Polystyrene is used for insulation and packaging.

Cambridge IGCSE Chemistry Study and Revision Guide © David Besser

Common error

- When asked to write an equation for the polymerisation of propene, a common response is

$$n\ CH_3CH{=}CH_2 \rightarrow -(CH_3CH{-}CH_2{-})_n$$

This is incorrect. The carbon atom in the CH_3 group has five bonds and the carbon atom in the CH_2 group has three bonds.

The monomer must be drawn as a displayed (or partially displayed) structure. The correct answer is shown in Figure 15.7.

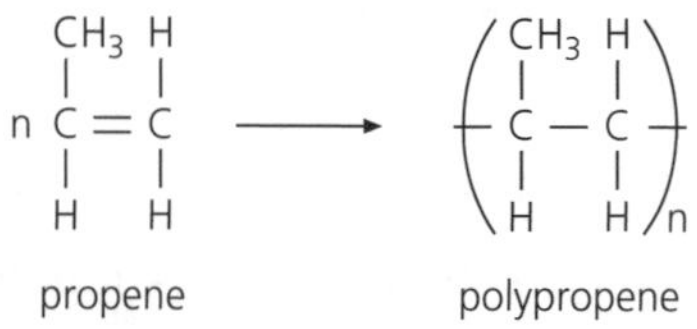

$$n\ \underset{propene}{\overset{CH_3\ \ H}{\underset{H\ \ \ \ H}{C=C}}} \longrightarrow \underset{polypropene}{\left(\overset{CH_3\ \ H}{\underset{H\ \ \ \ H}{-C-C-}} \right)_n}$$

Figure 15.7

Condensation polymerisation

A condensation reaction is a reaction in which two molecules join together and a simple molecule, such as water, is removed at the same time. Esterification is an example of a condensation reaction.

Condensation polymers are formed from monomers with two functional groups each. A simple molecule, such as water, is eliminated as the monomers join together.

Examples of such functional groups are —OH, —COOH and —NH_2.

Polyesters and polyamides are examples of condensation polymers.

Polyesters

Polyesters can be made from dicarboxylic acids (molecules with two —COOH groups) and diols (molecules with two —OH groups). These monomers can be represented as

HOOC—�—COOH   HO—▯—OH
   a dicarboxylic acid   a diol

Figure 15.8

The polymerisation occurs by the removal of a molecule of water when a —COOH group and an —OH group react. The monomers join together as

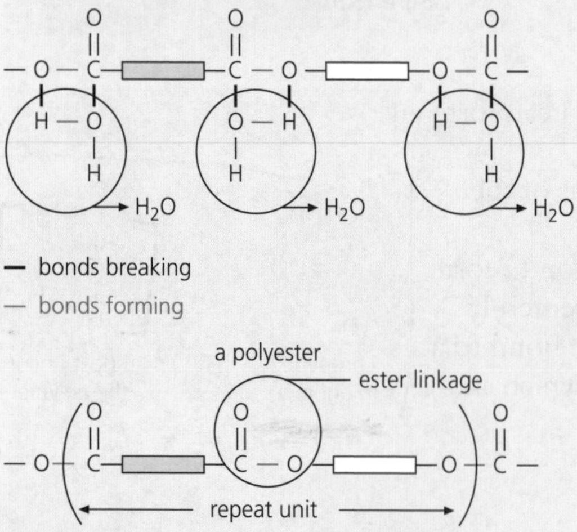

— bonds breaking
— bonds forming

Figure 15.9

Because there are —COOH groups and —OH groups at both ends of the monomers, more linkages can form and the polymer chain can grow in both directions.

Cambridge IGCSE Chemistry Study and Revision Guide © David Besser

Use of Terylene

Terylene is an example of polyester made from a dicarboxylic acid and a diol. It is a synthetic fibre used in clothing manufacture.

Polyesters can also be made using one monomer with both an —OH and a —COOH group, for example,

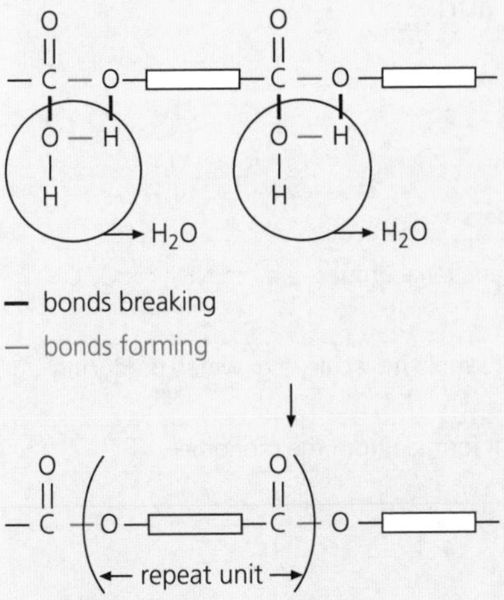

Figure 15.10

In this case, the repeat unit shows the residue of the single monomer molecule.

Polyamides

Polyamides can be made from dicarboxylic acids (molecules with two —COOH groups) and diamines (molecules with two —NH$_2$ groups). These monomers can be represented as

Figure 15.11

The polymerisation occurs by the removal of a molecule of water when a —COOH group and a —NH$_2$ group react. The monomers join together as

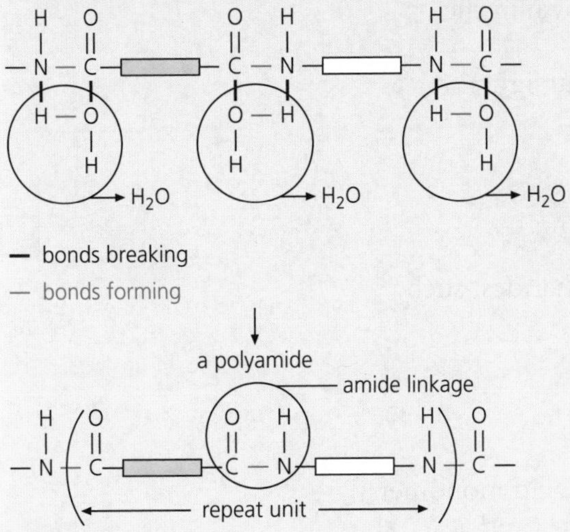

Figure 15.12

Because there are —COOH groups and —NH$_2$ groups at both ends of the monomers, more linkages can form and the polymer chain can grow in both directions.

Cambridge IGCSE Chemistry Study and Revision Guide © David Besser

Use of nylon

Nylon is an example of a polyamide made from a dicarboxylic acid and a diamine.

Nylon is a synthetic fibre used in clothing manufacture. It is also used in ropes, parachutes and strings for tennis rackets.

Polyamides can also be made using one monomer with both an —NH_2 and a —COOH group.

Table 15.5 Addition and condensation polymers

	Addition polymers	**Condensation polymers**
Monomers	Contain a C≡C double bond	Contain two reactive functional groups, e.g. —NH_2, —COOH, —OH
Polymerisation	Occurs without any loss of atoms producing only one product (the polymer)	Occurs with removal of simple molecule, e.g. water, producing two products
Polymers	Have same empirical formula as the monomer	Have different empirical formula from the monomers

Disposal of polymers

Household waste contains large quantities of polymers (plastic objects). These can be disposed of by

- burying in landfill sites
- incineration (burning).

Both methods of disposal contribute significantly to environmental pollution.

- Burying in landfill sites means that plastics will remain in the environment and take up large amounts of space, especially if they are non-biodegradable (which addition polymers are).
- Incineration can lead to production of toxic gases, such as carbon monoxide, and acidic gases, such as hydrogen chloride, which contribute to acid rain.

Attempts to overcome these problems include
- development of biodegradable plastics (those that break down in the environment as a result of bacterial activity)
- development of photodegradable plastics (break down in sunlight)
- sorting and recycling schemes.

● Natural polymers

Foods contain proteins and complex carbohydrates (polysaccharides, such as starch).

Proteins

Proteins are natural polyamides which are made from amino acid monomers. There are 20 different amino acids. All have an —NH_2 (amine) and a —COOH (carboxylic acid) functional group. These groups react together by condensation polymerisation to produce proteins which have amino acid residues in a sequence which is specific to each individual protein.

Cambridge IGCSE Chemistry Study and Revision Guide © David Besser

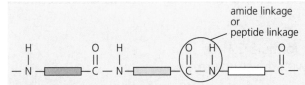

Figure 15.13 A protein molecule

Proteins contain the same amide linkage as that present in synthetic polyamides, such as nylon, although biologists usually refer to it as a peptide linkage when it exists in proteins.

Complex carbohydrates

Complex carbohydrates, such as starch, are naturally occurring condensation polymers made when glucose monomers

HO —⬜— OH
glucose

Figure 15.14 A glucose monomer

join together with the removal of water molecules to form starch.

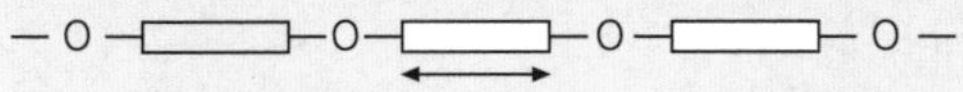

— O —⬜— O —⬜— O —⬜— O —

Figure 15.15 Starch

Hydrolysis of natural polymers

Proteins

The constituent amino acids that are present in different proteins can be identified by

- heating for 24 hours with 6.0 mol dm^{-3} (concentrated) hydrochloric acid. The protein undergoes acid catalysed hydrolysis (break down by reaction with water) and is converted to its constituent amino acids. (This is the reverse of condensation polymerisation.)

- separating the amino acids by paper chromatography (see Chapter 2)

- spraying the chromatogram (chromatography paper) with a locating agent (ninhydrin is used), so that the colourless amino acids produce visible blue spots

- determining R_f values and comparing with R_f values of known amino acids in a data book (see Chapter 2).

Complex carbohydrates

Complex carbohydrates can also be hydrolysed by hydrochloric acid. This breaks the complex carbohydrate down to simple sugars. The complex carbohydrate starch would be broken down into glucose by hydrolysis.

Hydrolysis can also be catalysed by enzymes. Starch is broken down by the enzyme amylase in saliva to produce maltose (a disaccharide made from two glucose units).

Exam-style questions

1 Three compounds, A, B and C, all have the molecular formula C_3H_8O. A reacts with ethanoic acid to produce a compound with the structure shown below.

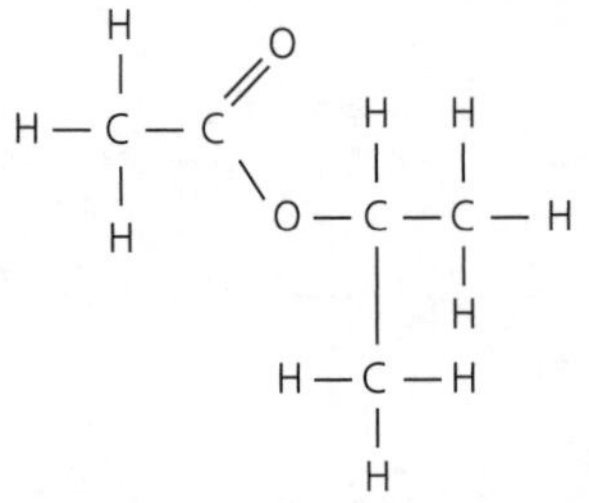

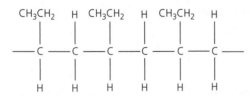

Figure 15.16

B reacts with ethanoic acid to produce a compound with the molecular formula $C_5H_{10}O_2$.

C does not react with ethanoic acid, but it undergoes complete combustion when burned in excess oxygen.

a What is meant by the term 'molecular formula'? [1 mark]

b What is the term used to describe compounds with the same molecular formula but different structural formulae? [1 mark]

c What is the empirical formula of the compound with the molecular formula $C_5H_{10}O_2$? [1 mark]

d What type of reaction occurs between A and ethanoic acid? [1 mark]

e What are the conditions that are required for A to react with ethanoic acid? [2 marks]

f Complete the chemical equation for the reaction occurring when C undergoes complete combustion in excess oxygen. State symbols are not required. [2 marks]

$$C_3H_8O + O_2 \rightarrow$$

g Give the structures of molecules A, B and C. Show all the atoms and all the bonds. [3 marks]
[Total: 11 marks]

2 The diagram shows part of a polymer.

$$\begin{array}{cccccc} CH_3CH_2 & H & CH_3CH_2 & H & CH_3CH_2 & H \\ | & | & | & | & | & | \\ -C- & C- & C- & C- & C- & C- \\ | & | & | & | & | & | \\ H & H & H & H & H & H \end{array}$$

Figure 15.17

a What type of a polymer is shown? [1 mark]

b Draw a circle around one repeat unit of the polymer. [1 mark]

c i Draw the structure of one molecule of the monomer. Show all the atoms and all the bonds.

ii Name the monomer. [2 marks]
[Total: 4 marks]

Cambridge IGCSE Chemistry Study and Revision Guide © David Besser

3 The diagram shows part of a polymer which is formed by condensation polymerisation.

$$ -\overset{\displaystyle O}{\overset{\displaystyle \|}{C}} - C_6H_4 - \overset{\displaystyle O}{\overset{\displaystyle \|}{C}} - \overset{\displaystyle H}{\overset{\displaystyle |}{N}} - C_6H_4 - \overset{\displaystyle H}{\overset{\displaystyle |}{N}} - \overset{\displaystyle O}{\overset{\displaystyle \|}{C}} - C_6H_4 - \overset{\displaystyle O}{\overset{\displaystyle \|}{C}} - \overset{\displaystyle H}{\overset{\displaystyle |}{N}} - H $$

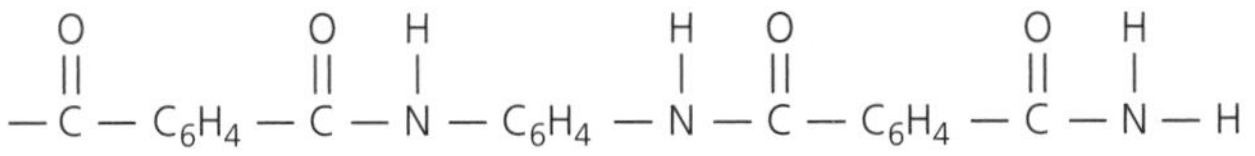

Figure 15.18

a What is meant by condensation polymerisation?   [2 marks]

b What type of condensation polymer is shown?   [1 mark]

c Draw a circle around one repeat unit of the polymer. Label the repeat unit.   [1 mark]

d Draw a circle around the linkage in the polymer. Label the linkage.   [1 mark]

e What type of biological molecule contains the same linkage as the polymer shown?   [1 mark]

f Draw the structures of the two monomers, showing all the atoms and bonds in the functional groups. (You may leave C_6H_4 as it is written.)   [2 marks]
                [Total: 8 marks]

4 The diagram shows part of a condensation polymer made from one monomer.

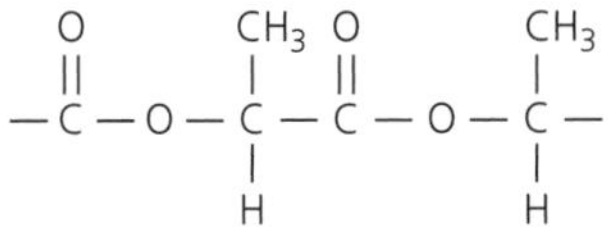

$$ -\overset{\displaystyle O}{\overset{\displaystyle \|}{C}} - O - \overset{\displaystyle CH_3}{\underset{\displaystyle H}{\overset{\displaystyle |}{\underset{\displaystyle |}{C}}}} - \overset{\displaystyle O}{\overset{\displaystyle \|}{C}} - O - \overset{\displaystyle CH_3}{\underset{\displaystyle H}{\overset{\displaystyle |}{\underset{\displaystyle |}{C}}}} - $$

Figure 15.19

a What type of a condensation polymer is drawn?   [1 mark]

b Draw a circle around the repeat unit of the polymer.   [1 mark]

c Draw the structure of the monomer showing all atoms and bonds.   [1 mark]

d What are the names of the two functional groups in the monomer?   [2 marks]
                [Total: 5 marks]

Key objectives

By the end of this section, you should be able to

- describe tests to identify
  - aqueous cations: aluminium, ammonium, calcium, chromium(III), copper(II), iron(II), iron(III), and zinc
  - cations: use of flame tests to identify lithium, sodium, potassium, copper(II)

- anions: chloride, bromide, iodide, carbonate, sulfite, sulfate and nitrate
- gases: ammonia, carbon dioxide, oxygen, hydrogen, chlorine and sulfur dioxide.

● Testing for cations (positive ions)

Cations (positive ions) can be identified using

- aqueous sodium hydroxide
- aqueous ammonia
- flame tests.

Using aqueous sodium hydroxide

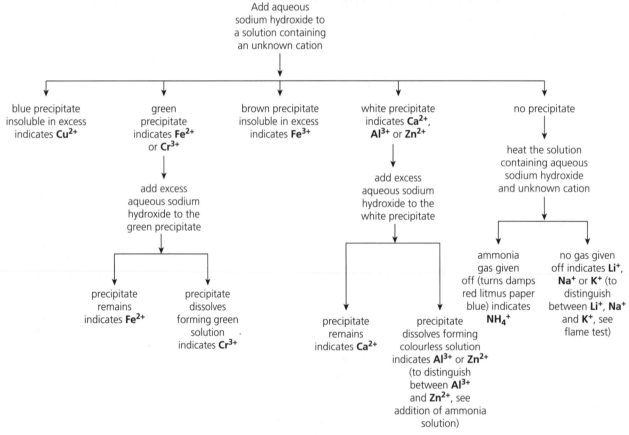

Figure 16.1 Testing for cations (positive ions) in aqueous solution using aqueous sodium hydroxide

Using aqueous ammonia

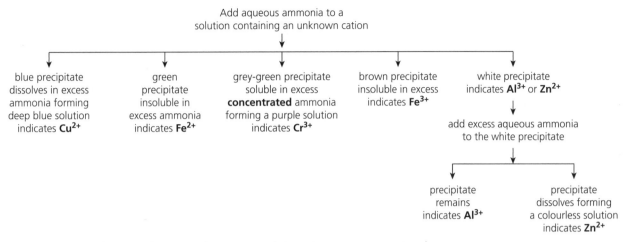

Figure 16.2 Testing for cations (positive ions) in aqueous solution using aqueous ammonia

Flame tests

Flame tests can be carried out on solids or on aqueous solutions.

- Starting with a solid, a few drops of concentrated hydrochloric acid should be added to a sample of the solid on a watch glass. Dilute hydrochloric acid can be used to avoid safety issues.
- A small amount of the mixture should then be placed on a nichrome wire.
- The nichrome wire containing some of the mixture is then placed in the hot part of a Bunsen flame.
- The colour of the flame identifies the cation (positive ion) (Table 16.1).

Table 16.1

Cation (positive ion)	Flame colour
Lithium, Li$^+$	Red
Sodium, Na$^+$	Yellow
Potassium, K$^+$	Lilac
Copper(II), Cu^{2+}	Blue-green

● Testing for anions (negative ions)

Tests for anions and their results are shown in Table 16.2.

Table 16.2

Test	Result	Anion
Add dilute nitric acid followed by aqueous silver nitrate	White precipitate	Chloride, Cl$^-$
	Cream precipitate	Bromide, Br$^-$
	Yellow precipitate	Iodide, I$^-$
Add any dilute acid	Bubbles Gas given off turns limewater milky (gas is CO$_2$)	Carbonate, CO$_3^{2-}$
	Gas given off when warmed Gas turns acidified aqueous potassium manganate(VII) colourless (gas is SO$_2$)	Sulfite, SO$_3^{2-}$
Add dilute nitric acid followed by aqueous barium nitrate	White precipitate	Sulfate, SO$_4^{2-}$
Add aqueous sodium hydroxide followed by aluminium. Warm gently	Gas given off turns damp red litmus paper blue (gas is NH$_3$)	Nitrate, NO$_3^-$

● Testing for gases

Tests for gases and their results are shown in Table 16.3.

Table 16.3

Test	Result	Gas
Damp red litmus	Turns blue	Ammonia, NH$_3$
Limewater	Turns milky	Carbon dioxide, CO$_2$
Glowing splint	Lights	Oxygen, O$_2$
Burning splint	Pops	Hydrogen, H$_2$
Damp litmus paper	Bleached	Chlorine, Cl$_2$
Acidified aqueous potassium manganate(VII)	Changes from purple to colourless	Sulfur dioxide, SO$_2$

Cambridge IGCSE Chemistry Study and Revision Guide © David Besser

Answers to exam-style questions

Chapter 1

1 Solid   [1]
Both melting point and boiling point are above 50°C.   [1]
Although the melting point being above 50°C is really enough of an explanation, it is advisable to refer to both melting point and boiling point.

2 **a** C   [1]
Solids have particles that are very close together, ordered and vibrate about a fixed position.

   b B   [1]
Liquids have particles that are fairly close together, irregularly arranged and move slowly.

   c D   [1]
Gases have particles very far apart, arranged randomly and moving at high speeds.

   d A   [1]
Particles that are very far apart must be in a gas which means they cannot be ordered or vibrate about fixed positions.
The question does not ask for explanations, so it is unnecessary to give any.

Chapter 2

1 **a** Dissolving (sugar in water)   [1]
Filtration (to remove sand)   [1]
Crystallisation   [1]

   b Distillation   [1]
(Simple) distillation, rather than fractional distillation is used to separate a pure liquid from a solution.

   c Fractional distillation   [1]
Fractional distillation is used to separate a mixture of two or more liquids with different boiling points. (Simple) distillation is incorrect.

   d Filtration   [1]
Washing with distilled water   [1]
Drying on a warm windowsill or in a low oven   [1]
A precipitate is an insoluble/undissolved solid. Centrifugation is another alternative. Neither of these processes will produce a pure solid, because the solid will be contaminated with a small amount of the solution it was separated from. Therefore, washing with distilled water and drying are both required in addition.

2 Heat until crystals form on a glass rod placed in the solution and withdrawn.   [1]
Leave the hot saturated solution to cool slowly. [1] Crystals should then form.
Remove crystals (by filtration if there is any liquid left).   [1]
Wash with a small amount of cold distilled water and then dry in a low oven or on a warm windowsill.   [1]
Common mistakes include
- *Not stating when the heating should be stopped.*
- *Evaporating to dryness. In this case water of crystallisation would be removed, leaving the anhydrous salt.*
- *Not specifying that a small amount of cold distilled water be used for washing. There is a danger of dissolving the crystals as well as removing impurities if too much is used or if the water is not cold.*
- *If an oven is used for drying it should be low. Too much heat would cause the crystals to decompose (particularly those that contain water of crystallisation).*

3 Carry out paper chromatography using a suitable solvent.   [1]
Allow solvent to reach the top of the chromatography paper.   [1]
Remove chromatography paper and allow to dry.   [1]
Spray with locating agent.   [1]
Measure R_f values and compare with data book values to identify amino acids.   [1]
A simple diagram of how paper chromatography is carried out is the best way to describe how the apparatus is set up even if the question does not ask for a diagram.
It is a common error to draw the solvent level above the starting line on the chromatography paper.

Chapter 3

1

Particle	Number of protons	Number of electrons	Electronic configuration	Charge on particle
A			2,8,8 [1]	2+ [1]
B				1− [1]
C	10 [1]			
D				2− [1]

Cambridge IGCSE Chemistry Study and Revision Guide © David Besser

2 a

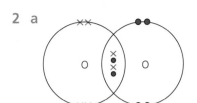

b

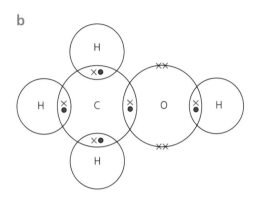

c

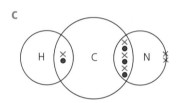

d

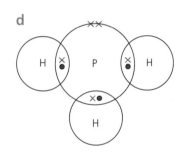

e

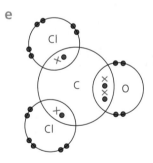

[Total: 5 marks]

Students are encouraged to use the type of diagrams shown. In many cases it is very difficult to decide which electrons are shared if diagrams are badly drawn.

'Outer shell electrons' only are requested.

3 a $Mg(OH)_2$
b $CaCl_2$
c $(NH_4)_3PO_4$
d Li_2S
e $Pb(NO_3)_2$
f $CaCO_3$
g $Al(NO_3)_3$

h K_2SO_3
i $ZnSO_4$
j $(NH_4)_2SO_4$ [1 mark each]
[Total: 10 marks]

Please note: (d) asks for a sulfide and (h) asks for a sulfite, whereas both (i) and (j) ask for a sulfate.

4 a B, C, D and E
Solids have melting points and boiling points above 25°C.

b A
A has a melting point below 25°C and a boiling point above 25°C so it is a liquid at 25°C.

c F
F has both melting point and boiling point below 25°C so it is a gas at 25°C.

d D
D is the only one to conduct electricity when solid; therefore D could have a giant metallic structure. (It could also be graphite or graphene.)

e C
C has a high melting point and a high boiling point and conducts electricity when molten but not when solid. Therefore C has a giant ionic structure.

f E
E has a high melting point and a high boiling point which means it has a giant structure.
E is a non-conductor both when solid (so it is not a giant metallic structure) and when molten (so it is not a giant ionic structure). Therefore E has a giant molecular structure.

[1 mark each]
[Total: 6 marks]

● Chapter 4

1 M_r of $H_2 = (1 \times 2) = 2$
Moles of Al $= 8.1 \div 27 = 0.3$ moles   [1]
Mole ratio from the equation
2 mole Al : 3 mole H_2
0.30 moles Al : $0.30 \times 3/2 = 0.45$ moles H_2 [1]
Mass of H_2
$=$ moles $\times$ mass of 1 mole
$= 0.45 \times 2 = 0.90$ g   [1]

[Total: 3 marks]

2 M_r of $KO_2 = 39 + (16 \times 2) = 71$
Moles of $KO_2 = 0.142 \div 71 = 0.002$ moles   [1]
Mole ratio from the equation
4 mole KO_2 : 3 mole O_2
0.002 moles KO_2 : $0.002 \times 3/4 = 0.0015$ moles of O_2 [1]
Volume of $O_{2(g)}$
$=$ moles $\times$ volume of one mole of gas
$= 0.0015 \times 24 = 0.036$ dm³   [1]

[Total: 3 marks]

Cambridge IGCSE Chemistry Study and Revision Guide © David Besser

3 M_r of $CaC_2 = 40 + (12 \times 2) = 64$

Moles of $C_2H_{2(g)} = 120 \div 24\,000 = 0.005$ moles $C_2H_{2(g)}$ [1]

(The volumes must both be in the same units: see below.)

Because the volume in the question is given in cm³, the volume of 1 mole of a gas must be converted from 24 dm³ to 24 000 cm³. Alternatively, 120 cm³ could be converted to 0.120 dm³.

Mole ratio from the equation

1 mole C_2H_2 : 1 mole CaC_2

0.005 moles C_2H_2 : 0.005 moles CaC_2 [1]

Mass of CaC_2

= moles × mass of 1 mole

= $0.005 \times 64 = 0.32$ g [1]

4 Moles of $H_2SO_4 = \dfrac{35.0 \times 0.20}{1000} = 0.007$ [1]

Mole ratio in equation

1 mole H_2SO_4 : 2 moles KOH

0.007 moles H_2SO_4 : $2 \times 0.007 = 0.014$ moles KOH [1]

a concentration of KOH

$\dfrac{\text{moles} \times 1000}{\text{volume (cm}^3)} = \dfrac{0.014 \times 1000}{20.0}$

= 0.70 mol dm⁻³ [1]

b M_r of KOH = $39 + 16 + 1 = 56$

To convert concentration in mol dm⁻³ to concentration in g dm⁻³

Mass (grams) = moles × M_r

= 0.70×56

= 39.2 g dm⁻³ [1] [Total: 4 marks]

5 a Calculate the number of moles of atoms of each element.

Carbon, C $54.5 \div 12 = 4.54$

Hydrogen, H $9.1 \div 1 = 9.1$

Oxygen, O $36.4 \div 16 = 2.275$

Divide all the above by the smallest

C $4.54 \div 2.275 = 2$

H $9.1 \div 2.275 = 4$

O $2.275 \div 2.275 = 1$

Empirical formula = C_2H_4O

b If the M_r of the compound = 44

The M_r of $C_2H_4O = (12 \times 2) + (1 \times 4) + 16 = 44$

n = M_r of the compound ÷ M_r of empirical formula

$n = 44 \div 44 = 1$

Therefore, molecular formula = $C_2H_4O \times 1 = C_2H_4O$

6 M_r of $TiCl_4 = 48 + (35.5 \times 4) = 190$

Moles of $TiCl_4 = 0.38 \div 190 = 0.002$ moles [1]

Mole ratio 1 mole of $TiCl_4$: 1 mole of Ti

Therefore, 0.002 moles of $TiCl_4$: 0.002 moles of Ti [1]

Mass of Ti = $0.002 \times 48 = 0.096$ g [1]

0.096 g is 100%. But the yield is only 0.024 g.

Percentage yield = actual yield ÷ 100% yield × 100%

Percentage yield = $0.024 \div 0.096 \times 100 = 25.0\%$ [1]

[Total: 4 marks]

It is essential to show all working out in all calculations.

● Chapter 5

1

Electrolyte	Name of product at anode (+)	Name of product at cathode (−)
Molten potassium bromide	Bromine [1]	Potassium [1]
Aqueous potassium bromide	Bromine [1]	Hydrogen [1]
Molten lead iodide	Iodine [1]	Lead [1]
Aqueous copper(II) chloride	Chlorine [1]	Copper [1]
Aqueous sodium sulfate	Oxygen [1]	Hydrogen [1]

Molten electrolytes can only produce the two elements present in the molten compound. Aqueous electrolytes produce
- *halogens or oxygen at the anode (exceptions apply when the anode is made of an unreactive metal such as copper)*
- *metals below hydrogen in the reactivity series or hydrogen at the cathode.*

Metals above hydrogen in the reactivity series are never produced by electrolysis of aqueous solutions.

2 a Nickel is the anode. [1]

b A solution containing nickel ions is the electrolyte. [1]

c The object to be plated, i.e. the knife, is the cathode. [1]

Students may not be familiar with nickel compounds, but if asked to name a soluble compound it is safe to assume that nitrates are always soluble in water. Therefore aqueous nickel nitrate could be chosen as the electrolyte.

3 a Iodine [1]

Halogens are always produced at the anode from halides (in preference to oxygen).

b $2H^+ + 2e^- \rightarrow H_2$ [1]

Potassium is never produced by electrolysis of an aqueous solution.

It is unnecessary to give state symbols in equations unless requested.

c Oxidation [1]

Oxidation always occurs at the anode because electrons are always lost by ions being discharged at the anode.

d (Aqueous) potassium hydroxide   [1]

Hydrogen ions (from water) are discharged at the cathode and iodide ions are discharged at the cathode. This leaves potassium ions, K⁺, and hydroxide ions, OH⁻ (from water), in the solution, which becomes (aqueous) potassium hydroxide. This is similar to the electrolysis of concentrated aqueous sodium chloride in which hydrogen and chlorine are produced at the electrodes and the electrolyte becomes aqueous sodium hydroxide.

e Electrons   [1]

Conducting wires, being metallic, conduct electricity because they contain moving electrons. 'Ions' is a common incorrect answer.

f (K⁺ and I⁻) ions   [1]

Electrolytes conduct electricity because they contain moving ions.
'Electrons' is a common incorrect answer.

● Chapter 6

1 a

[2]

b 2 C–C, 8 C–H, 5 O = O [2]

*It is a common error to include **3** C–C bonds because the formula is C_3H_8.*

c $(2 \times 347) + (8 \times 435) + (5 \times 497) = 6659$ kJ [1]

d 6 C=O, 8 O–H [2]

e $(6 \times 803) + (8 \times 464) = 8530$ kJ   [1]

f $8530 - 6659 = 1871$ kJ/mole. The reaction is exothermic.   [3]

Because the amount of energy given out when new bonds form in the products (8530) is bigger than the amount of energy taken in to break bonds in the reactants (6659), the overall energy change is exothermic.

● Chapter 7

1 a Physical   [1]

This produces aqueous sodium chloride. The sodium chloride and water are not changed into any substances that are not there at the start. Since a new chemical substance is not produced it is a physical change.

b Chemical   [1]

This produces three new chemical substances, i.e. chlorine, hydrogen and sodium hydroxide.

Therefore, a chemical change occurs.

c Chemical   [1]

The silver chloride is chemically changed into silver and chlorine.

d Physical   [1]

This is a method of separation and does not produce any new chemical substances.

e Physical   [1]

This is a method of separation and does not produce any new chemical substances.

2

Experiment	Hydrochloric acid	Calcium carbonate	Temperature/°C	Graph
1	50 cm³ of 0.10 mol dm⁻³	Marble chips	25	A
2	50 cm³ of 0.20 mol dm⁻³	Marble chips	25	D [1]
3	50 cm³ of 0.10 mol dm⁻³	Powdered	25	B [1]
4	50 cm³ of 0.10 mol dm⁻³	Marble chips	12.5	E [1]
5	50 cm³ of 0.10 mol dm⁻³	Marble chips	50	B [1]

The volume of hydrogen produced depends only on the number of moles of hydrochloric acid used (because the calcium carbonate is in excess). Experiment 2 is the only experiment in which more moles of hydrochloric acid are produced. The number of moles of hydrochloric acid is doubled and so the volume of hydrogen is doubled. Because the concentration of hydrochloric acid is greater, the rate of reaction is faster and the graph is steeper (Graph D). Increasing the surface area (Experiment 3) and increasing the temperature (Experiment 5) both increase the rate of reaction without changing the volume of hydrogen. Thus the graph is steeper in both cases and levels off at the same volume (Graph B). In Experiment 4, decreasing the temperature decreases the rate and the graph is less steep but levels off at the same volume because the volume of hydrogen is unchanged (Graph E).

3 a There are the same numbers of gas molecules (2) on both sides of the ⇌ sign; therefore increasing the pressure does not favour either reaction.   [1]

b There are two molecules of gas on the right and three molecules of gas on the left of the ⇌ sign; therefore increasing the pressure causes equilibrium to shift in the direction of fewer gas molecules, which means to the right.   [1]

c There are two molecules of gas on the right and one molecule of gas on the left of the ⇌ sign; therefore increasing the pressure causes equilibrium to shift in the direction of fewer gas molecules, which means to the left.   [1]

115

4 a The forward reaction is exothermic, which means that the reverse reaction is endothermic. When temperature is decreased, equilibrium always shifts in the exothermic direction. In this case, equilibrium shifts to the right.   [1]

b The forward reaction is endothermic, which means that the reverse reaction is exothermic. When temperature is decreased equilibrium always shifts in the exothermic direction. In this case, equilibrium shifts to the left.   [1]

5 a $Mg_{(s)} + Cu^{2+}_{(aq)} \rightarrow Mg^{2+}_{(aq)} + Cu_{(s)}$ [1]
The sulfate ions, $So_4{}^{2+}$, are spectator ions. State symbols need not be given in equations unless asked for.

b Oxidation: $Mg_{(s)} \rightarrow Mg^{2+}_{(aq)} + 2e^-$ [1]
Reduction: $Cu^{2+}_{(aq)} + 2e^- \rightarrow Cu_{(s)}$ [1]
The total number of charges on both sides of any equation must be equal.

c Cu^{2+} [1] is an electron acceptor/gains electrons.   [1]
Oxidising agents are always reduced.
Oxidising agents are always on the same side as the electrons in an ionic half-equation.

d Mg [1] is an electron donor/loses electrons.   [1]
Reducing agents are always oxidised.
Reducing agents are always on the opposite side to the electrons in an ionic half-equation.

Chapter 8

1 a i Method 1
ii (Dilute) hydrochloric acid   [1]
Hydrochloric acid produces chlorides.
iii $CoCO_3 + 2HCl \rightarrow CoCl_2 + CO_2 + H_2O$
[1 formulae, 1 balancing]
Cobalt compounds are not all that well known, but the (II) in cobalt(II) carbonate and cobalt(II) chloride mean that a cobalt ion is Co^{2+}, thus making it straightforward to derive the formulae of the compounds.

b i Method 3
ii (Aqueous) sodium iodide   [1]
To decide on a choice of an aqueous solution containing iodide ions, you should remember that all sodium (or potassium) salts are soluble in water.
iii $2NaI + Pb(NO_3)_2 \rightarrow PbI_2 + 2NaNO_3$
[1 formulae, 1 balancing]

c i Method 2   [1]
ii (Dilute) nitric acid   [1]
Nitric acid produces nitrates.
iii $KOH + HNO_3 \rightarrow KNO_3 + H_2O$
[1 formulae, 1 balancing]

2 • Pour dilute sulfuric acid [1] into a beaker.
• Add magnesium carbonate.   [1]
• Stir or warm.   [1]
• Stop adding magnesium carbonate when some remains undissolved/no more bubbles of gas evolved.   [1]
• Filter off excess magnesium carbonate.   [1]
• Heat the filtrate until crystals form on a glass rod placed in the solution and withdrawn.   [1]
• Leave the hot saturated solution to cool slowly. [1] Crystals should then form.
• Remove crystals (by filtration if there is any liquid left).   [1]
• Wash with a small amount of cold distilled water and then dry in a low oven or on a warm windowsill.   [1]
• $MgCO_3 + H_2SO_4 \rightarrow MgSO_4 + CO_2 + H_2O$ [1]

3 • Add aqueous sodium hydroxide or potassium hydroxide.   [1]
• Stir/warm to dissolve scandium oxide.   [1]
• Filter off copper(II) oxide.   [1]
• Wash with distilled water.   [1]
• Dry the copper(II) oxide on a warm windowsill/in a low oven.   [1]

Chapter 9

1 a B [1]
Group I elements become more reactive down the group.
b D [1]
Group VII elements become more reactive up the group.
c Either F or G   [1]
d A [1]
Periods are the horizontal rows.
e A [1]
Atoms of Group IV elements have four electrons in their outer shells. All atoms have the same number of outer electrons as their group numbers.

2 a Bubbles, or lithium floats, or lithium disappears, or lithium melts, or lithium moves around
Any three [3]
It is a common error to say lithium bursts into flame. Only the elements from potassium downwards burst into flame.
b $2Li_{(s)} + 2H_2O_{(l)} \rightarrow 2LiOH_{(aq)} + H_{2(g)}$
[1 formulae, 1 balancing, 1 state symbols]
This equation is very commonly asked for (with any Group I element) in exams. The formulae and balancing numbers are the same for all Group I elements; only the symbol of the element changes.

Cambridge IGCSE Chemistry Study and Revision Guide © David Besser

c Yellow [1]

The colours of methyl orange and litmus in acidic and alkaline solutions should be memorised.

3 a $Cl_{2(g)} + 2KI_{(aq)} \rightarrow 2KCl_{(aq)} + I_{2(aq)}$
[1 formulae, 1 balancing, 1 state symbols]
$Cl_{2(g)} + 2I^-_{(aq)} \rightarrow 2Cl^-_{(aq)} + I_2$ [1]

b $Br_{2(l)} + 2KI_{(aq)} \rightarrow 2KBr_{(aq)} + I_{2(aq)}$
[1 formulae, 1 balancing, 1 state symbols]
$Br_{2(l)} + 2I^-_{(aq)} \rightarrow 2Br^-_{(aq)} + I_{2(aq)}$ [1]

$I_{2(s)}$ is acceptable in both cases because some of it may form as a precipitate.

The halogen elements are often dissolved in water to carry out these reactions, in which case (aq) would be used as the state symbol.

Potassium ions are spectator ions in both reactions. If aqueous sodium iodide was used in either or both cases, the sodium ions would be spectator ions and the ionic equations would be exactly the same.

4 a $F_2 + 2KCl \rightarrow 2KF + Cl_2$
[1 formulae, 1 balancing]

b No reaction [1]

c $Br_2 + 2KAt \rightarrow 2KBr + At_2$
[1 formulae, 1 balancing]

d No reaction [1]

Halogens nearer the top of Group VII displace those lower down but not vice versa.

Reactions involving fluorine only occur in theory, because in practice fluorine reacts violently with water so could not be used.

Astatine is radioactive.

5 a Cu_2O [1]

b $Cu(NO_3)_2$ [1]

c $FeCl_2$ [1]

d $Fe_2(SO_4)_3$ [1]

The roman numerals used to represent oxidation states are the same as the number of positive charges on the cations: copper(I) refers to Cu^+, iron(II) refers to Fe^{2+} and iron(III) refers to Fe^{3+}.

● Chapter 10

1 a B, C, A, D [1]

b $B + A(NO_3)_2 \rightarrow A + B(NO_3)_2$ [1]

c $C + D^{2+} \rightarrow D + C^{2+}$ [1]

Because the nitrate ion is NO_3^-, it can be deduced that all the charges on the metal ions are 2+ because 1 metal ion is combined with 2 nitrate ions in each compound as can be seen from the formulae.

d B [1]

The most reactive metal is always the negative terminal in a cell containing two dissimilar metals in an electrolyte, because the most reactive metal

has the greatest tendency to release electrons (which makes it negative) as it forms positive ions.

e $B + DO \rightarrow BO + D$ [1]

Because B and D have ions with the formula B^{2+} and D^{2+}, and because the oxide ion is O^{2-}, the formulae of the metal oxides are BO and DO.

2 ● Add a dilute acid in excess (preferably hydrochloric or sulfuric but not nitric). [1]

● Stir or warm or both.

● Zinc reacts and dissolves but copper does neither; bubbling is also seen with zinc but not with copper. [1]

● Filter off copper. [1]

● Wash with distilled water and dry in a low oven. [1] [Total: 4 marks]

Students should memorise the position in the reactivity series of those elements referred to on the syllabus. Zinc is above hydrogen but copper is below hydrogen. Thus zinc reacts with dilute hydrochloric and sulfuric acids to produce an aqueous salt solution. Hydrogen gas is evolved.

3

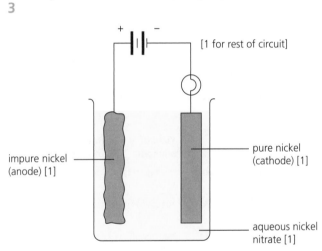

[1 for rest of circuit]

impure nickel (anode) [1]

pure nickel (cathode) [1]

aqueous nickel nitrate [1]

[Total: 4 marks]

Because both electrodes contain nickel, the anode must be clearly labelled as impure nickel and the cathode as pure nickel. Students should remember that all nitrates are soluble in water; therefore aqueous nickel nitrate (or nickel nitrate solution) is a suitable electrolyte.

4 a B, copper(II) nitrate [1]
C, copper(II) oxide [1]
D, nitrogen dioxide [1]
E, oxygen [1]

Students are asked to name the substances. It is a common error to write formulae.

b i $CuO + 2HNO_3 \rightarrow Cu(NO_3)_2 + H_2O$
[1 formulae, 1 balancing]

ii $2Cu(NO_3)_2 \rightarrow 2CuO + 4NO_2 + O_2$
[1 formulae, 1 balancing]

c Copper(II) hydroxide [1]
Copper(II) carbonate [1]

Cambridge IGCSE Chemistry Study and Revision Guide © David Besser

Bases (copper(II) hydroxide) and carbonates (copper(II) carbonate) are used to react with dilute acids to form aqueous solutions of soluble salts. Copper metal is not suitable because copper is below hydrogen in the reactivity series and therefore it does not react with dilute acids.

Chapter 11

1 a Brown [1] to black [1]
   A colour change is requested; therefore it is essential to give the original and final colours.

   b $2Cu + O_2 \rightarrow 2CuO$ [1 formulae, 1 balancing]

   c Oxygen gas is taken out/removed from the air. [1]

   d 'Excess' in this case means that there is more than enough copper to react with all the oxygen. Some copper will be left over when all the oxygen has reacted. [1]
   Students commonly answer that excess means 'more than enough' but it is important to add more than enough to react. The point is emphasised by saying some will be left over after the reaction.

   e Nitrogen [1]

2 a A catalyst increases the rate of a chemical reaction [1] and is chemically unchanged at the end of the reaction. [1]

   b

	Yield of ammonia	Rate of production of ammonia
Higher temperature than 450 °C	Decrease [1]	Increase [1]
Higher pressure than 200 atmospheres	Increase [1]	Increase [1]
Not using a catalyst	No change [1]	Decrease [1]

   The table makes distinctions between yield (equilibrium position) and rate of reaction. Students should make sure that they do not confuse one with the other (see Chapter 7). Where there are changes the answers must be comparative, e.g. lower yield and not low yield.

   c i Sulfuric acid [1]
      ii $2NH_3 + H_2SO_4 \rightarrow (NH_4)_2SO_4$ [1 formulae, 1 balancing]

   d $3NH_3 + H_3PO_4 \rightarrow (NH_4)_3PO_4$ [1 formula of $(NH_4)_3PO_4$, 1 balancing]

Chapter 12

1 a Vanadium(V) oxide or vanadium pentoxide [1]
   The oxidation state of vanadium is an essential part of the name vanadium(V) oxide. Iron (the catalyst in the Haber process) is a common error.

   b Below 450 °C, the rate of reaction would be slower. [1]

   c The rate of reaction would increase at a higher pressure [1] because the gas molecules would be closer together and therefore there would be more collisions in any given time. [1]
   Stating that gas molecules would be closer together therefore there would be more collisions is not enough for the second mark. Reference to time is essential.

   d The yield of sulfur trioxide and the rate of reaction would be high enough at 1–2 atmospheres.
   or
   There would be risks of explosions/leakages at higher pressure.
   or
   It would be too expensive to build an industrial plant to withstand higher pressure. Any two [2]

   e Sulfur trioxide is dissolved in 98% concentrated sulfuric acid to produce oleum. $SO_3 + H_2SO_4 \rightarrow H_2S_2O_7$ [1]
   The oleum is then added to the correct amount of water to produce sulfuric acid of the required concentration.
   $H_2S_2O_7 + H_2O \rightarrow 2H_2SO_4$ [1]

2 a $H_2SO_4 + CuCO_3 \rightarrow CuSO_4 + CO_2 + H_2O$ [1 formulae of $CuCO_3$ and $CuSO_4$, 1 rest of equation completely correct]

   b $H_2SO_4 + 2NaOH \rightarrow Na_2SO_4 + 2H_2O$ [1 formulae, 1 balancing]

   c $H_2SO_4 + KOH \rightarrow KHSO_4 + H_2O$ [1]

   d $H_2SO_4 + Zn \rightarrow ZnSO_4 + H_2$ [1]

Chapter 13

1 Plants grow better in soils at specific pH values. [1]
   Calcium oxide or calcium hydroxide [1]
   Lime, quicklime or slaked lime are common names (as also are limestone and lime water) and not chemical names.

2 a i $2C_8H_{18} + 25O_2 \rightarrow 16CO_2 + 18H_2O$ [1 formulae, 1 balancing]
      Fractions/multiples are acceptable in equations unless stated otherwise.

      ii $CaCO_3 \rightarrow CaO + CO_2$ [1]

      iii $Fe_2O_3 + 3CO \rightarrow 2Fe + 3CO_2$ [1 formulae, 1 balancing]

      iv $CaCO_3 + 2HCl \rightarrow CaCl_2 + CO_2 + H_2O$ [1 formulae, 1 balancing]

   b i Water [1]
      ii Glucose [1], oxygen [1]
      iii UV light/sunlight [1], chlorophyll [1]

Chapter 14

1

[structural formulae diagrams]

[2]

When drawing molecules it is essential that all carbon atoms have four bonds only.

It is also essential to draw two different molecules, and not draw the same molecule twice as in the following examples.

[structural formulae diagrams]

These are all diagrams of the same molecule.

They are not structural isomers.

These show the same molecule drawn several different ways. In all the examples, the double bond is between the first two carbon atoms and all five carbon atoms are in a 'straight chain' even if they are drawn at an angle in some cases.

2 a Ultraviolet light   [1]
   b Chloroethane   [1]
   c Any one of the following:
      $C_2H_4Cl_2 / C_2H_3Cl_3 / C_2H_2Cl_4 / C_2HCl_5 / C_2Cl_6$
      [1]

   Some or all of the hydrogen atoms in ethane can be substituted by chlorine atoms. The total number of atoms bonded to the two carbon atoms must be six in each case.

3 a Addition   [1]
   b The bromine changes colour from yellow/brown [1] to colourless.   [1]

   When the bromine test for unsaturation is described, it is essential to give the initial and final colours.

If excess propene is not used there may be some bromine left and so the colour of the bromine may still be visible.

   c i $C_3H_6Br_2$ [1]
      Two bromine atoms are added to the propene.
     ii C_3H_8 [1]
      Two hydrogen atoms are added to the propene.
    iii C_3H_8O [1]
      A water molecule is added to the propene, i.e. two hydrogen atoms and an oxygen atom are added.
   d Nickel   [1]

4 a C_9H_{20} [1]
      The correct answer is obtained using n = 9 in the general formula for alkanes which is C_nH_{2n+2}.
   b Any one from the following:

[structural formula]   or $CH_3CH_2CH = CH_2$

but-1-ene

[structural formula]   or $CH_3CH = CHCH_3$

but-2-ene

[structural formula]   or $CH_3 - C = CH_2$ with CH_3

2-methylpropene

[formula 1, name 1]
      Allow structural formulae or displayed formulae because 'showing all the atoms and bonds' is not requested. Structural formulae must show the C=C double bond.
   c i $C_8H_{18} \rightarrow C_4H_{10} + 2C_2H_4$ [1]
     ii $C_8H_{18} \rightarrow C_5H_{10} + C_3H_6 + H_2$
         or $C_8H_{18} \rightarrow C_2H_4 + C_6H_{12} + H_2$
         or $C_8H_{18} \rightarrow C_2H_4 + 2C_3H_6 + H_2$ [1]
      There are acceptable answers with an alkyne C_nH_{2n-2} and an alkane as the products as well as hydrogen.

Chapter 15

1 a The formula that shows the number of atoms of each element in one molecule of an element or compound.   [1]
   b Structural isomers   [1]
      The word 'structural' is often omitted.
   c $C_5H_{10}O_2$ [1]
      This formula cannot be simplified any further. Therefore the molecular formula and the empirical formula are the same.

d Esterification   [1]
Condensation would also be acceptable.

e Heat [1] and a catalyst of concentrated sulfuric acid   [1]

f $2C_3H_8O + 9O_2 \rightarrow 6CO_2 + 8H_2O$
[1 formulae, 1 balancing]
Fractions/multiples are accepted in equations.

g

A

H   H   H
| | |
H—C—C—C—H
| | |
H   O   H
    |
    H

B

H   H   H
| | |
H—C—C—C—O—H
| | |
H   H   H

C

H   H   H
| | |
H—C—C—O—C—H
| | |
H   H   H
                [3]

The diagram below shows parts of the esters formed from A and B.

```
    O                         O
    ‖                         ‖
 —C                        —C
    \   H   H                 \   H   H   H
     O—C²—C¹                   O—C¹—C²—C³—H
         |   |                     |   |   |
         H                         H   H   H
     H—C³—H
         |
         H
```

part of the ester   part of the ester
formed from A   formed from B

The oxygen atom in the ester formed from A is joined to carbon number 2. Therefore the O–H group in alcohol A must have been on carbon number 2.

The oxygen atom in the ester formed from B is joined to carbon number 1. Therefore the O–H group in alcohol B must have been on carbon number 1.

C does not react with ethanoic acid which means that C is not an alcohol and does not have an O–H group. The only molecule that can be drawn with the molecular formula C_3H_8O that is not an alcohol and obeys the rules that
* *carbon atoms have only four bonds*
* *oxygen atoms have only two bonds, and*
* *hydrogen atoms have only one bond is*

H   H   H
| | |
H—C—C—O—C—H
| | |
H   H   H

2 a Addition polymer   [1]

b

```
 /CH₃CH₂  H\CH₃CH₂  H  CH₃CH₂  H
|   |    |    |    |    |    |
—C — C  — C — C  — C — C —
|   |    |    |    |    |    |
 \  H    H/   H    H    H    H
```
                [1]

The circle should be drawn around two consecutive carbon atoms in the main chain and all the atoms and groups of atoms joined to them. One example is given.

c i

```
    H   H       H
    |   |       |
H — C — C — C = C
    |   |   |   |
    H   H   H   H
```
                [1]

ii But-1-ene   [1]

3 a Condensation polymerisation is the formation of a long-chain molecule (the polymer) from small molecules (monomers) [1]. A simple molecule such as water is eliminated as the monomers join together.   [1]

b A polyamide   [1]

c, d

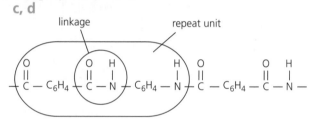

                [1 each]

e A protein   [1]

f

```
H               H               O               O
|               |               ‖               ‖
N — C₆H₄ — N    H — O — C — C₆H₄ — C — O — H
|               |
H               H
```
                [2]

In carboxylic acids and alcohols, it is a common error for students not to draw the bond between O and H atoms, i.e. they draw –OH instead of –O–H.

4 a A polyester   [1]

b

```
    O         CH₃  O           CH₃
    ‖         |    ‖           |
 —C + O — C — C + O — C —
              |                |
              H                H
```
                [1]

c

```
        H
        |
    H — C — H   O
        |       ‖
    H — C —— C — O — H
        |
        O
        |
        H
```
                [1]

d Carboxylic acid   [1]
  Alcohol   [1]

Cambridge IGCSE Chemistry Study and Revision Guide © David Besser

Index

Cambridge IGCSE Chemistry Study and Revision Guide © David Besser

Cambridge IGCSE Chemistry Study and Revision Guide © David Besser